T0290864

Recent Advances in Energy Harvesting Technologies

RIVER PUBLISHERS SERIES IN ENERGY SUSTAINABILITY AND EFFICIENCY

Series Editors:

PEDRAM ASEF
Lecturer (Asst. Prof.) in Automotive Engineering, University of Hertfordshire, UK

The "River Publishers Series in Sustainability and Efficiency" is a series of comprehensive academic and professional books which focus on theory and applications in sustainable and efficient energy solutions. The books serve as a multi-disciplinary resource linking sustainable energy and society, fulfilling the rapidly growing worldwide interest in energy solutions. All fields of possible sustainable energy solutions and applications are addressed, not only from a technical point of view, but also from economic, social, political, and financial aspects. Books published in the series include research monographs, edited volumes, handbooks and textbooks. They provide professionals, researchers, educators, and advanced students in the field with an invaluable insight into the latest research and developments.

Topics covered in the series include, but are not limited to:

- Sustainable energy development and management;
- Alternate and renewable energies;
- Energy conservation;
- Energy efficiency;
- Carbon reduction;
- Environment.

For a list of other books in this series, visit www.riverpublishers.com

Recent Advances in Energy Harvesting Technologies

Editors

Shailendra Rajput

Xi'an International University, China

Abhishek Sharma

Graphic Era University, India

Vibhu Jately

University of Petroleum and Energy Studies, India

Mangey Ram

Graphic Era University, India

River Publishers

Routledge
Taylor & Francis Group

NEW YORK AND LONDON

Published 2023 by River Publishers
River Publishers
Alsbjergvej 10, 9260 Gistrup, Denmark
www.riverpublishers.com

Distributed exclusively by Routledge
605 Third Avenue, New York, NY 10017, USA
4 Park Square, Milton Park, Abingdon, Oxon OX14 4RN

Recent Advances in Energy Harvesting Technologies / by Shailendra Rajput, Abhishek Sharma, Vibhu Jately, Mangey Ram.

© 2023 River Publishers. All rights reserved. No part of this publication may be reproduced, stored in a retrieval systems, or transmitted in any form or by any means, mechanical, photocopying, recording or otherwise, without prior written permission of the publishers.

Routledge is an imprint of the Taylor & Francis Group, an informa business

ISBN 978-87-7022-845-9 (print)
ISBN 978-87-7022-993-7 (paperback)
ISBN 978-10-0096-283-3 (online)
ISBN 978-1-003-44038-3 (ebook master)

While every effort is made to provide dependable information, the publisher, authors, and editors cannot be held responsible for any errors or omissions.

Contents

8 Energy Harvesting from Conducting Nanocomposites   177

Ankit Kumar Srivastava, Swati Saxena, Sonika, and
Sushil Kumar Verma

Preface

The growing need for energy by the human race calls for the search for sustainable resources to ensure energy security. Once the resources are identified, the effort has to be directed to harvest as much energy from these as is technologically feasible and economically viable.

Harvesting energy can be broadly divided into two levels: at the macro level and at the micro level. Technologies for harvesting energy would also need to be grouped into two corresponding groups. In the first group at the macro level, the focus primarily is on meeting the gross energy requirements of a region, state or country from the considerations of energy security as well as sustainability. Reduction of carbon emissions remains an important factor too while deciding the resources to be harvested and the technology used in the harvesting of energy. Extracting maximum power from solar, wind etc falls in this category. So do all technologies that assist in maintaining the grid stable and secure. All major energy storage devices and systems with their associated techniques fall into this group too. In fact, technologies for sustainable agriculture including irrigation and hydrological studies also fall into this category.

The second group at the micro level has a wide arrange of technologies that can help us in harvesting small amounts of energy through miniature, wearable and portable energy sensing elements like piezoelectric nano-generators (PENG), triboelectric nano-generators (TENG), nano-composite electrical generators (NEG), thermoelectric generator (TEG), etc. These mini harvesters can be used for harvesting the small amounts of energy associated with human movement, like walking, stepping, dancing, etc., that can be useful for self-powering small smart devices. The flexible and transparent harvesters made of nanocomposites can help in supplementing the battery life if not able to replace the battery altogether. Some polymer nanocomposites have shown promise for building thermoelectric coolers that can be used for transporting vaccines and serums. Because of the diffused form of small amounts of these harvesters, the role of AI and machine learning techniques can be handy in making the systems scalable and economically viable.

Inspiration for compiling this book basically came from the need felt by the editors to have in a single volume, the *Recent Advances in Energy Harvesting Technologies*. We are confident that this volume containing contributions from authors on energy harvesting technologies at the macro level in Chapters 2, 3, 4, 7 and 9 and at the technologies at the micro level in the remaining chapters will be found interesting and of value by the readers.

Editors:

Shailendra Rajput
Xi'an International University, China

Abhishek Sharma
Graphic Era University, India

Vibhu Jately
University of Petroleum and Energy Studies, India

Mangey Ram
Graphic Era University, India

Acknowledgement

The editor acknowledges River publishers for this opportunity and professional support. My special thanks to Ms Philippa Jefferies, River publishers for the excellent support, she provided us to complete this book.

Thanks to the chapter authors and reviewers for their availability for this work.

List of Figures

List of Tables

List of Contributors

Ashok, Prashasti, *Department of Geology, Institute of Earth Sciences, India*

Azzopardi, Brian, *MCAST Energy Research Group (MCAST Energy), Institute of Engineering and Transport, Malta College of Arts, Science and Technology (MCAST), Malta*

Bisht, Neeraj, *Department of Mechanical Engineering, College of Technology, GBPUAT, India*

Chauhan, Sakshi, *Department of Mechanical Engineering, College of Technology, GBPUAT, India*

Chinthamalla, Ramulu, *EE Department, National Institute of Technology, India*

Dutt, Sunil, *Department of Chemistry, Government Post Graduate College, India*

Gopikishan Sabavath, *Faculty of Science, Department of Physics, University of Allahabad, India*

Gorfad, Vijaykumar, *Department of Aerospace Engineering, MIT School of Engineering, MIT ADT University, India*

Hong Lim, Wei, *Faculty of Engineering, Technology and Built Environment, UCSI University, Malaysia*

Hu, Xinghao, *Institute of Intelligent Flexible Mechatronics, Jiangsu University, China*

Jadhav, Vishwanath, *Department of R&D (NPD), Deep Plast Industries, India*

Jately, Vibhu, *Department of Electrical and Electronics Engineering, University of Petroleum and Energy Studies, India*

Joshi, Jyoti, *Department of Computer Science and Engineering, Graphic Era Hill University, India*

Krishna Mathi, Dileep, *EEE Department, Vidya Jyothi Institute of Technology, India*

Kumar Bajaj, Dinesh, *MIT-ADT University, India*

Kumar Srivastava, Ankit, *Department of Physics, Indrashil University, India*

Kumar Verma, Sushil, *Department of Chemical Engineering, Indian Institute of Technology Guwahati, India*

Pal Singh, Surendra, *Surveying Engineering Department, Wollega University, Ethiopia*

Sahoo, Devabrata, *Department of Aerospace Engineering, MIT School of Engineering, MIT ADT University, India*

Saxena, Swati, *Department of Physics, Sardar Vallabhbhai National Institute of Technology, India*

Sharma, Abhishek, *Department of Computer Science and Engineering, Graphic Era Deemed to be University, Dehradun, India*

Siddharth, Pandi, *MIT-ADT University, India*

Sonika, *Department of Physics, Rajiv Gandhi University, Doimukh, Itanagar, India*

Tashale Amanu, **Marga,** *Surveying Engineering Department, Wollega University, Ethiopia*

List of Abbreviations

ABC	Artificial bee colony
AI	Artificial intelligence
ANN	artificial neural network
AR	Augmented reality
CANN	cluster-ANN
CGSVM	Coarse-Gaussian support vector machine
CNN	Convolution Neural Network
CNTs	Carbon nanotubes
ELM	Extreme learning machine
EU	European Union
EVs	Electric vehicles
FOA	Firefly optimization algorithm
GMPP	Global MPP
GP	Global peak
HBB-BC	Hybrid big bang-big crunch
IGSA	Improved version of the gravitational search algorithm
ILC	Iterative learning controller
INC	Incremental conductance
IoT	Internet of Things
I–V	Current versus voltage
ML	Machine learning
MLR	Multiple linear regression
MPP	Maximum power point
MPPT	Maximum power point tracking
P&O	Perturb and observe
P-ANN	Periodic-ANN
PSC	Partial shading conditions
PV	Photovoltaic
P–V	Power versus voltage
RNN	Recurrent neural network
SAE	Stacked auto-encoder

SI	Swarm intelligence
SOA	Seeker optimization algorithm
SVM	Support vector machine
T-ANN	Threshold-ANN
UI	User interface
UIC	Uniform irradiance conditions
VR	Virtual reality

1

AI in Energy Harvesting

Vijaykumar Gorfad[1], Devabrata Sahoo[1], and Xinghao Hu[2]

[1]Department of Aerospace Engineering, MIT School of Engineering, MIT ADT University, India
[2]Institute of Intelligent Flexible Mechatronics, Jiangsu University, China
E-mail: gorfadvijay@gmail.com; devabrata.sahoo@mituniversity.edu.in; huxh@ujs.edu.cn

Abstract

Humankind consumes different forms of energy for life and growth. Electric energy is one of them. Demand for electric energy keeps growing as the application area of electric energy keeps increasing day by day. To meet this huge demand and to save natural recourses, there is research going on to generate electrical energy from renewable energy resources like solar energy, wind energy, etc. Energy harvesting technology is an emerging area that gives the opportunity to increase the efficiency of reusable energy generation by using recent technology like artificial intelligence, machine learning, the Internet of Things, etc. The dissipated energies like electromagnetic waves, heat energy and vibrations are converted to electric energy. Artificial intelligence can be deployed along with different sensors like piezoelectric sensors for energy harvesting. This harvested energy can effectively be useful in different sectors like automobiles, domestic and industrial applications, etc. This book chapter summarizes the characteristics of energy harvesting with the help of AI/ML, recent developments in this area and major challenges, along with the future scope of development. Application of existing artificial intelligence technology in the field of energy harvesting is explained and the possible use of existing state-of-the-art AI/ML technologies for prediction and increase the efficiency of energy harvesting technology is also suggested in the present book chapter.

Keywords: Energy Harvesting; AI/ML; Piezoelectric Energy Harvester; Electro-magnetic Energy Harvester; Triboelectric energy harvester (TENG).

1.1 Introduction

There is unlimited energy available in nature surrounding us. A small amount of this energy can be harvested and used as electrical energy. This energy harvested from nature plays important role in achieving sustainable development goals. Nowadays, with the increasing use of technology, the requirement for energy is further increasing. To fulfil these energy requirements, effective implementation of energy harvesting became the need of the day. Energy harvesting is mainly dealing with photovoltaics in the visible light range. The process of energy harvesting can be classified mainly into four processes: energy harvesting from the environment, conversion of this energy into electric energy, power conservation and power management. Figure 1.1 shows this process in detail.

Solar cells are one of the most widely used technologies for energy harvesting for a long time. Apart from sunlight, researchers are also focusing on the stable output from indoor light whose spectrum is spread in the narrow bend of the visible light. There are different types of solar cells used for energy harvesting, there is a need to have a standard process to evaluate energy harvesting characteristics of the solar cells to implement those technologies in society and carry out more research and development activities on energy harvesting.

Solar power technology is described in various research articles and in high-level review papers by a number of researchers. This chapter is mainly focused on three techniques for collecting vibration, radio waves and thermo-electric energy. The parameters mainly needed for power harvesting devices applied to Internet of Things instruments are tiny and high-performance. Also, environmental resistance and functional dependability requirements depend on the application environment, execution type and economical involvement. As depicted in Figure 1.1, the energy harvesting technology is assumed to be applied to IoT fields such as clearing the information and transmission of sensor data and excludes geothermal, wave and wind power generation as well as large-scale environmental energy harvesters.

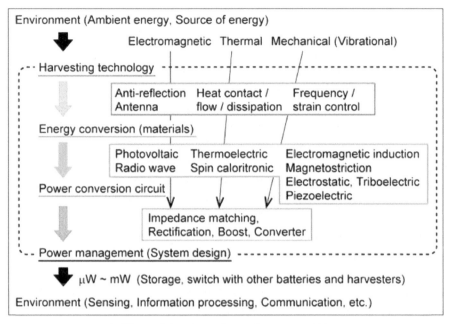

Figure 1.1 Energy harvesting process [1].

1.2 Energy from Mechanical Vibrations

Mechanical vibrations can be used as a source of energy to be harvested. The vibration energy harvester senses the mechanical vibration in the surroundings and converts this vibration energy into electrical energy which can be used for remote or wireless applications. As this energy is of low level it can be used for electronic devices such as sensors whose energy requirement is low specifically in terms of mW. Vibrational energy can be harvested using the following three methods:

1. electromagnetic,
2. electrostatic and
3. piezoelectric.

In the **electromagnetic method** of vibrational energy harvesting, electromagnetic induction and inverse magnetostrictive effects are used. In this method, permanent magnets are used to apply the bias magnetic field to control the magnetisation state of the magnetostrictive material. Afterward, the variation in magnetic flux is achieved by using a strain. This variable magnetic flux is converted into electric energy with the help of coils. Electret

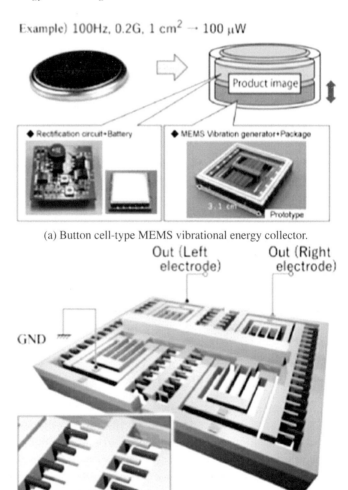

Example) 100Hz, 0.2G, 1 cm^2 → 100 μW

(a) Button cell-type MEMS vibrational energy collector.

(b) Button cell-type MEMS vibrational generator and package.

Figure 1.2   MEMS vibrational energy harvester of button cell type [1].

vibrational energy harvesting with the help of MEMS and triboelectric energy harvesting comes under the **electrostatic method** of vibrational energy harvesting. In this method, the electric energy is generated with the help of fluctuating electrostatic capacitance. The fluctuations in the electrostatic capacitance are produced with the help of the vibrations of the electrode of a capacitor. Piezoelectric energy is generated when piezoelectric materials come in contact with the vibrations. When mechanical stress or strain is applied to the dielectric surface charge produced, this effect is known as the piezoelectric effect. The **piezoelectric method** uses this phenomenon to harvest energy from the vibrations. The performance of all three methods depends on the type of vibrations and the frequencies of the vibrations. These harvesters can work for the vibration of up to 200 Hz. If we take the example of vibrations of the human body or some of the infrastructures such as buildings or bridges, the frequency of these vibrations is approximately 2–3 Hz with an acceleration of approximately 10 m/s^2. Also, the vibration in the real condition is random in nature so as does the frequency and the accelerations. The performance of these energy harvesters is mainly depended upon the characteristics of these harvesters, such as output power, impedance and frequency response. The designs of the vibrational energy harvesters are, therefore mainly focused on impedance matching and energy conversion rate by avoiding the resonant conditions. Current research is focused on the development of harvesters to efficiently generate electric energy from MEMS which are being used for smaller vibrations such as with an acceleration of 0.98 m/s^2 and frequency of less than 100 Hz with the help of MEMS. Figure 1.2 shows a schematic diagram of a button cell-type MEMS vibrational energy collector. When the electret of one of the opposing comb electrode pairs is charged, the current is produced by electrostatic induction which produces vibrations in the electrodes. The comb electrodes make the device bulkier as the effective area is increased due to its structure. To overcome these problems, the use of advanced MEMS is essential at this point of time. CMOS integrated circuits are used to make voltage booster rectifiers which can give an output voltage of approximately 3.3 V. This low output of the rectifier helps to increase the frequency band by approximately 10 times in comparison to p-n rectifiers.

1.3 Fundamentals of Vibrational Energy Harvesting

The vibration energy harvesting system works on the basis of a seismic mass's resonance operation which electromechanically converts kinetic energy into electric energy. Only in a limited natural frequency bandwidth of the resonant frequency machine can the vibration energy harvester operate effectively. The resonance mechanism is an essential component of any vibration energy harvesting mechanism. The seismic mass oscillations within the mechanism get converted according to the physical principle of electromechanical transformation. Vibration energy harvesting system typically employs piezoelectric, electrostatic, or electro-magnetic conversion principles. Several techniques, such as spring non-linearity, can be used to extend the operating frequency's narrow bandwidth. Another critical parameter of the vibration energy harvester is the quality factor of the resonance mechanism. In resonant mode, the maximum amplitude of the relative motion of the resonant mechanism is estimated by the quality factor, which determines the collected power. The electromagnetic vibration energy converter explained here is having a magnetic circuit and a fixed air coil.

Figure 1.3 depicts the energy harvesting system developed by Hadas and his team. When this system is excited by vibration, the resonance mechanism's design causes vibrations in the magnetic circuit which is having a permanent magnet around the fixed coil. The oscillatory movement of the magnetic field induces a voltage in the coil due to Faraday's law. The use of harvested power in an electrical load results in energy dissipation. This means that an electrical load connected to the circuit will produce an electric current. According to Faraday's law, the oscillating motion of a magnetic field induces a voltage in the coil. Consumption of power collected from electrical loads causes energy loss. This means that the connected electrical load provides a current to the electrical circuit, producing feedback proportional to the electrical energy collected in the resonant circuit. The vibration energy rapidly decreases as the electric energy is harvested and with this, the amplitude of the vibration also reduces in the resonance circuit. This results in an induced voltage that is much lower than that of an open electrical circuit.

Vibration energy harvesting systems produce maximum power at the time of equivalent mechanical and electrical damping. This phenomenon is popular and has been inferred in several instances (Williams and Yates

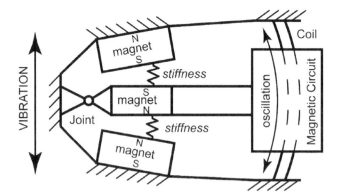

Figure 1.3   Springless electromagnetic vibration energy harvester [2].

1996; Hadas et al. 2010a). This fact must be taken into account whenever a vibrational energy harvesting model is developed, after which an optimal vibrational energy harvesting system is developed. Though the system can be again optimised to minimise volume and weight, or it can be optimised for maximum power output. The main factors of these studies are the quality factors and mass of the resonant mechanism, the geometry, the parameters of the electromagnetic transducer, the electrical load, the materials and technology used, etc. Hadas et al developed the vibration energy harvester under the European project WISE, which was again modified to maximise vibration sensitivity and the efficiency of the system in terms of power output. This device works on the electromagnetic principle of conversion of mechanical to electrical energy. This system converts electromagnetic energy into electric energy [2]. Optimisation is carried out for maximum power output and to minimise the volume and weight of the device. Artificial intelligence methods can be used to optimise the parameters of the harvester device. The authors applied AI techniques to optimise the mechanical, electrical and electromagnetic permeates of the harvester device.

1.4 Piezoelectric and Triboelectric Nanogenerators

Piezoelectric nanogenerators (PENG) and triboelectric nanogenerators (TENG) are energy harvesting technologies that harvest energy from environments using piezoelectric and triboelectric effects, respectively. In the last 10 years, both of these technologies are growing rapidly and they have reignited the interest among researchers towards green energy and energy harvesting. At a time of the considerable power shortage and remarkable ecological downturn caused by conventional fossil fuel consumption, it becomes important to restructure the energy generation system from an unreproducible conventional energy generation method to ecologically sound reproducible energy harvesting technologies such as PENG and TENG. PENG and TENG are to outperform traditional power solutions because of their efficiency, potential for producing power from different natural scenarios like body movements, fluctuating temperatures, structural oscillations, etc., and out-of-the-box electromechanical characteristics.

Because of the characteristics of piezoelectric and triboelectric materials, the structural and material elements play an important role in these energy harvesting technologies; so it is important to logically design the structure and artificial materials for getting the improved electrical properties that are generally absent in the conventional structures or conventional materials. As a result, a new research focus has been shifted to the design, characterisation and use of PENGs and TENGs to get the desired electrical performance which results in a wide interdisciplinary research area that includes power, electricity, structural analysis, materials science and multiscale commerce, manufacturing and multi-functional applications. Energy harvesting devices are reported for use in a variety of usage apart from energy harvesting that includes new equipment and methods in engineering and biomedical fields. Due to the exponential growth of the recent information era, efficient power supplies and a wide range of energy solutions are urgently needed to charge different advanced commercial and information technologies like smart sensors and monitors, recognition and computing in high-tech cities. However, these energy harvesters are currently struggling to provide clean energy solutions in a more portable, dependable and eco-friendly way. As a result, research efforts have emerged to design, predict as well as optimise the performance of these devices using artificial intelligence (AI). AI, an intelligent machine that tries to imitate human perception, has been widely used as a supportive model for solving difficult engineering problems which conventional approaches cannot solve. Artificial intelligence has received considerable attention in the present time because of the inefficiency of conventional models developed using the first principles of physics. AI technologies improve computers' ability to solve problems by copying complicated bioactivities like learning, rational thinking and auto-correction. Early attempts identified artificial intelligence and its derivatives as an effective approach to solving the problems of designing, predicting and optimising structures and materials in energy harvesting techniques. Unlike conventional statistical methods, AI methods can capture precise operational relationships between input parameters like environmental conditions and exogenous variables like electrical energy in absence of prior assumptions of any relationships. The benefits of AI over conventional methods in PENG and TENG are primarily due to the characteristics that promote computational efficiency in designing new structures and discovering materials. Using AI to mine, process and analyse data has improved the precision and effectiveness of the modified structural design and piezoelectric and triboelectric material research in energy harvesting devices. As a result, the present model for deploying AI in harvesting devices has emerged. Although AIPENG and AITEN offer advanced mechanical and electrical performance, an overview of bright topics has not yet been explored.

1.5 Electro-mechanical Energy Harvesting

Mechanical to electrical energy harvesting is designed to produce the electricity from the environment in an eco-friendly, inexpensive, reproducible, dependable and greener way. In addition to the popular electromagnetic generators, various techniques have been designed to collect all sorts of energy like solar, nuclear, thermal, chemical, magnetic, etc. Mechanical energy is usually wasted due to its low amplitude, low frequency, low energy density and diffusion form. As a new green energy solution, piezoelectric and triboelectric nano harvesters are developed to harvest energy from mechanical energy sources which are constant, unconventional, easily procurable and extensive. PENGs and TENGs are piezoelectric and friction materials with an easy design that can be activated by mechanical excitation and produce power more efficiently. Mechanical–electrical energy harvesting is designed to harvest cheaper, renewable, green, energy from the environment. In addition to the popular electromagnetic generators, other methods like PENG and TENG were developed primarily based on the output current in Maxwell's theory of displacement current. Debuting in 2006, PENG uses the piezoelectric effect to convert strains or stresses caused by mechanical excitation into electrical energy. First reported in 2012, TENG uses the triboelectric effect to convert the combined effects of contact charging and electrostatic induction into electrical energy. Compared to their counterparts in the energy harvesting technology suite, PENG and TENG have demonstrated advantages, ranging from the conceptual economics of environmental power generation to the manufacturing and applicability of self-powered units on multiple scales.

Figure 1.4 (a) shows the fundamentals of the piezoelectric and triboelectric effects. The piezoelectric effect of PENG can be displayed in a metal–insulator–metal sandwich structure comprising an insulating piezoelectric sheet in-between two sheets of the metal electrode. Initially, positive and negative ions are superimposed, so there is no polarisation in the piezoelectric material. This reduces the volume of the piezoelectric material and creates negative stresses as a result of the deformation. The positive and negative ions separate to create an electric dipole and the electric dipole moment changes, creating a piezoelectric potential in-between the electrodes.

By attaching the electrodes to an outer excitation, the piezoelectric charge causes electrons to flow into an external circuit, partially shielding the potential and attaining a new equilibrium. This is how energy is harvested from mechanical vibrations. The maximum compression state with the maximum polarisation density is achieved when the two conductive electrodes

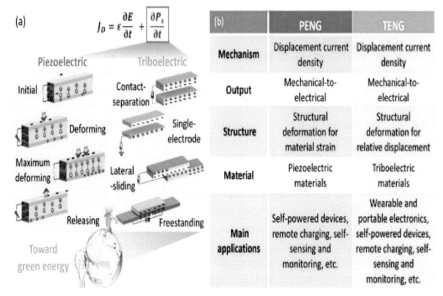

Figure 1.4   Comparisons of PENG and TENG [3].

come into full contact by means of deformation. Finally, when the external force is released from the shunt, the electrons return to a new equilibrium state. The triboelectric effect of a heating element is depending on the joint effect of contact charging and electrostatic induction. This nano harvester is considered as a multi-purpose mechanical energy harvesting technology that efficiently converts external mechanical energy into electrical energy. In general, the more two triboelectric materials are separated; the more electric charge can move along the triboelectric series, thus generating higher power. The triboelectric effect is widespread in a variety of materials, but triboelectric materials have a significant impact on charging efficiency and final power. Therefore, it is very important to choose a triboelectric material for the heating element. The selection of a set of triboelectric materials with opposite triboelectric polarities is the key to getting increased power. The mechanism, output, structure, material and applications of PENG and TENG are compared in Figure 1.4 (b). Depending on the bias current density mechanism, the PENG and TENG generate an electrical output from a mechanical input power source. Due to fundamental differences, PENG structures are usually designed to activate piezoelectric materials by material deformation whereas TENGs activate triboelectric materials by relative displacement. Thus, PENG and TENG have developed microstructures engineered to get unusual electromechanical behaviours in terms of design, such as PENGs reinforced with mechanical Meta materials, TENGs reinforced with spring or pivot structures, etc. In addition, PENG and TENG have been reported to provide a decisive increase in the energy density generated using functional materials in terms of material properties such as PZT-based compositions.

1.6 Piezoelectric Energy Harvesters

As per the piezoelectric effect, EEG generates electricity due to deformation due to external excitation. An electric dipole is created when an electric charge builds up due to strain stress at both ends of a piezoelectric material. The positive ion displacement of the anion leads to a piezoelectric potential. According to existing research, piezoelectric energy harvesting is an energy harvesting technology that can convert the mechanical energy of various scales into electrical energy. Various types of piezoelectric materials are developed as piezoelectric energy harvesters, which can be divided into piezoelectric ceramics and piezoelectric polymers. Piezoelectric ceramics are the semiconductor nanomaterials, lead-based ceramics and lead-free ceramics and piezoelectric polymers generally refer to organic polymers such as polyvinylidene fluoride (PVDF) and inorganic polymers. Piezoceramics are chemically inert, resistant to moisture and have elevated electromechanical conversion efficiencies, but their harsh properties make them insufficient for flexible device applications. In contrast, piezoelectric polymers have a little piezoelectric effect but are much more flexible. Research is done to explore the major properties of designing, predicting and optimising PENGs in response to various external influences, such as body movements or environmental influences. Figure 1.5 shows the existing PENG techniques.

Figure 1.5 (a) shows a MoS_2 piezoelectric energy harvester monolayer passivated with sulphur vacancy. Sulphur vacancies are passivated using a sulphur treatment on the original MoS_2 surface to expand the output peak current to 100 pA and voltage to 22 mV of a sulphur-treated MoS_2 piezoelectric energy harvester nano sheet monolayer. Sulphur treatment increased the maximum output by almost a factor of 10. The obtained data indicate that sulphur treatment can stop the screening effect and decreases the number of free charge carrier PENGs due to sulphur passivation. Figure 1.5 (b) shows a skin-like, battery-free piezoresistive sensor that can be conformal to human skin. The sensor was developed from four material groups, composed of (1) a multilayer NFC chip package, a loop antenna and a silicon pressure sensor, (2) a polyimide thin film as an electrical insulator and (3) a top and base.

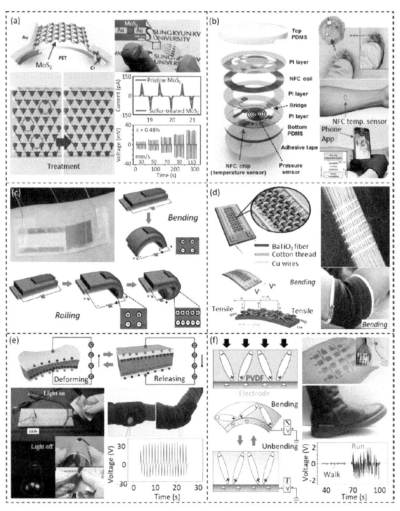

Figure 1.5 Existing piezoelectric energy harvesters [3]. (a) MoS2 piezoelectric energy harvester monolayer passivated with sulphur vacancy. (b) Battery-free piezoresistive sensor. (c) Solution-treated piezoelectric energy harvester. (d) 2D fabric PENG using PVC BaTiO3 nanowire hybrid piezoelectric fibres. (e) High tensile PZT-based PENG. (f) Electrode PENG with PVDF nano fibbers.

Polydimethylsiloxane as summarisation and (4) bio-friendly skin glue. The test results displayed the capabilities of the skin-mounted sensor by using a smart device to read data from a short distance. It is concluded that sensors may have promising value in monitoring circadian cycles and reducing risk.

Figure 1.5 (c) depicts solution-treated piezoelectric energy harvester flexible thin films by using drenched ZnO ink as a thin layer. Electrical energy is produced by the mechanical deformation of the elastic thin film during the rolling and bending process. It was reported that the ZnO PENG thin film showed high elasticity and mechanical fatigue resistance. a solution-treated p-type polymer blend and a perforated transport layer are placed to increase the power output. Figure 1.5 (d) shows a 2D fabric PENG using PVC BaTiO3 nanowire hybrid piezoelectric fibres. The piezoelectric properties of PVC BaTiO3 nanowires were enhanced by inorganic active BaTiO3 nanowires. The reported PENG was applied to an arm-actuated elbow pad, resulting in high performance and the ability to be used as a wearable power supply. Figure 1.5 (e) shows a high tensile PZT-based PENG that achieved an output power density of about 81 μW/cm^3. PZT particles are added to a solid silicone rubber with the help of the blending method. Therefore, the part of PZT in the composite rose from elongation of 30–92% by weight. This nano harvester is then connected to the human body to collect kinetic energy in multiple deformation modes. Figure 1.5 (f) depicts a single-electrode PENG with PVDF nano fibbers capable of performing pressure measurements with cold/heat integration with a single device. The piezoelectric signal is received as a square wave signal while the thermal sensor signal appears as a pulsed signal.

1.7 Triboelectric Energy Harvester (TENG)

According to the triboelectric effect, a heating element that generates energy from electric charges after contacting various triboelectric materials by friction has been developed. particularly, when two triboelectric materials are misshaped and brought into contact by an external excitation, an electrostatic charge is induced on the contact surface by the triboelectric effect. When these materials in contact are separated, more electric potentials are created between the electrostatic charges, moving the charges on the conductive material side by side. Utilising the triboelectric effect, TENG is used in multi-scale energy harvesting applications because of the properties such as environmental friendliness, more efficiency, less weight, cheapness and ease of availability. TENGs are generally classified into four categories of usage, like triboelectric modes of vertical contact separation, lateral slip, single electrode and freestanding triboelectric layers, making them versatile and versatile applications in various work conditions such as wastewater treatment, wireless communication, etc. Its possible application may be

Textiles, Human Movement, Human-Machine Interface (HMI), Vibration, Wind, Flowing Water, etc.

Figure 1.6 (a) shows a PETN-based micro-motion sensor operating through the fusion of triboelectricity and electrostatic induction. This elastic

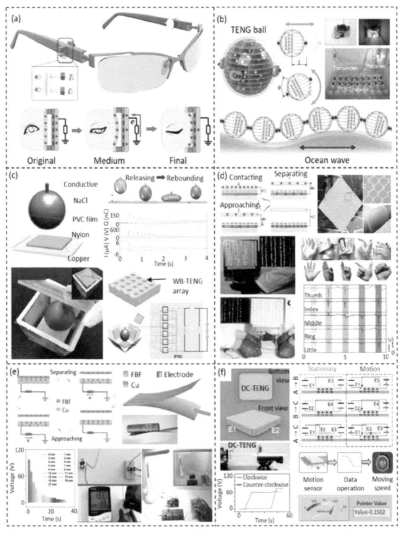

Figure 1.6   Existing TENG [3]. (a) PETN-based micro-motion sensors, (b) TENG for energy generation from water waves, (c) multiple-frequency TENG, (d) Flexible and Waterproof Skin-Inspired Piezoelectric Nanogenerator (PENG), (e) Adaptable, Exceptionally Flexible, and Highly Sensitive Triboelectric Nanogenerator (TENG) using Fish Bladder Films, (f) Self-sustained Triboelectric Vector Sensors using Direct Current TENG for Motion Parameters measurements.

and see-through sensor composed of an indium tin oxide electrode and two opposing friction materials successfully detected blinking motion by a strong signal level of about 750 mV. The sensor was attached to the goggles and applied to two real-time mechanosensory HMIs are highly sensitive, stable, economical and simple in operating. TENG-based micro-motion sensors were a smart measurement technology available on HMI. The macroscopic auto-assembly network composed of the entrapped TENG for energy generation from water waves is depicted in Figure 1.6 (b). Based on the self-adaptive magnetic joints of the encapsulated TENG, the network demonstrated self-assembly, self-healing and easy reconfiguration. The three-dimensional electrode structures are created to improve the TENG unit's output. Figure 1.6 (c) shows a multiple-frequency TENG based on a water balloon for energy harvesting in any direction of a water wave. Because of the excellent mechanical properties of the water balloon, the TENG achieved multiple frequency response using low-frequency external mechanical sim-ulations for generating high-frequency electrical of the instant short circuit output. In this method approximately the maximum instantaneous short-circuit current measured by the authors was 147 A, with an open-circuit high voltage.

1.8 Artificial Intelligent in Energy Harvesting

The foremost artificial intelligent technologies at present used in the energy harvesting sector are dealing with auto-gain of the information based on existing data, recognising invisible relationships in the inputs and outputs, and helping decision-making. The application of artificial intelligent for energy harvesting can be divided into information gathering and expression, algorithm definition and model construction as depicted in Figure 1.7 (a).

AI technologies are required to be trained for information gathering and plotting using available data of energy harvesting. Therefore, it is of utmost importance to maintain effective and sufficient data collection. In energy harvesting maintaining the capacity and efficiency of the data pool is critical as data collection and presentation will require training AI methods on existing energy harvesting data. For example, energy harvesting due to body motion is subjected to millions of periodic loads over a 24-hour period, typically resulting in 30% data loss which is the noise in data. Therefore, pre-processing of the initial information is required to be performed to understand missing or incorrect data during data collection.

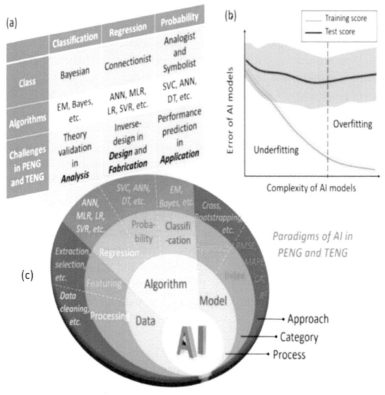

Figure 1.7   Artificial intelligence in energy harvesting [3]. (a) Artificial intelligent for energy harvesting. (b) The interrelation between the difficulties and errors of the AI model. (c) Artificial intelligence in energy harvesting.

Recognising and fixing these bugs is important to reduce the chance of an AI model being misguided. The large set of information in energy harvesting is so critical to the precision and effectiveness of AI models. For example, the voltage produced is recorded in the time domain. However, the output must be transformed into the frequency domain for more analysis. Like human intelligence, AI algorithms are far more efficient than remaining on some specific row information types. Transforming this data into a format appropriate for AI models requires functional characterisation or development. Converting the information into a suitable arrangement for AI models, characteristic study, or characteristic engineering is required to be carried out.

When defining an algorithm, it is necessary to define a specific algorithm for training a dataset after collecting and presenting the data. A wide pool of

AI algorithms can be run to construct the predictive energy harvesting models like supervised machine learning (ML) models utilised in forecasting energy generation depending on individual data sets that require categorisation of the data set or continuous data sets that require data regression. There can be simple and rigid or flexible and complex AI predictive models for energy harvesting. AI constructs the algorithms to increase model performance depending on big data by mimicking human learning abilities and gaining experience through thinking. Data is a key parameter in AI models, but big data is of no use until a computer can draw conclusions or knowledge. AI usually uses some assumptions and implements them in the model to effectively capture missing data. A normal AI paradigm may be described as the use of different methods to determine unique acts and relationships between resultant data and invisible patterns or features of materials and structures. In PENG and TENG, a grouping of algorithms can be used, inter alia, for analytical problems, regression algorithms for design and manufacturing problems and probabilistic algorithms for applied problems. During model construction, the models need to be evaluated, validated and optimised to obtain the best AI model for energy harvesting. The precision of the AI model can be tested using the cross-validation method if the inputs used for training and validation can represent the total data set, but problems can arise when the data set is smaller. Appropriate acts should be applied for testing the applicability of AI models. Different indicators (such as RMSE, MAPE, CA, R2, etc.) are suggested to test and analyse the projection errors of AI models. Model bias and difference are largely regarded as the main causes of model error in energy collection. Model bias refers to errors due to improper assumptions in an algorithm, resulting in a lack of a fundamental relationship between the input raw data and energy efficiency predictions. Model variance describes the sensitivity to small variations in the input raw data, such as data noise, computational uncertainty and measurement constraints. Relatively high bias or high variance usually degrades the performance of AI models. Figure 1.7 (b) shows the interrelation between the difficulties and errors of the AI model. High bias is considered underfitting, which happens at a time when an AI model is rigid enough to account for the difficult relationship between the input raw data (e.g., the power harvesting voltage observed in an experiment) and its predictive function (e.g., the forecasted voltage). The large variation is taken as overfitting that takes place when the AImodel is more complicated like increasing input variables.

Figure 1.8 (a) explains a typical structure of an artificial neural network (ANN) composed of multiple abstract brain neurons and data processing connections. Nodes represent a particular output performance and the relationship between nodes is the weights of the transform signal (i.e., the memory of the ANN model). As a result, the output of an ANN is strongly influenced by the weights, features and relationships of the model. ANN is used in AI algorithms in energy harvesting methods because of its high nonlinear approximation ability. Figure 1.8 (b) shows the structure of a support vector machine (SVM) that classifies input data using supervised learning. The nature of SVM as a generalised classifier is to determine a support vector for the formation of an optimal classification hyperplane on the training information volume. SVM uses a hinge loss function to estimate factual risk and therefore improves model soundness and slenderness. Figure 1.8 (c) shows the structure of an extreme learning machine (ELM) as a learning algorithm for a feed-forward neural network having one invisible surface. ELM can randomly generate network parameters. For example, the threshold of the hidden layer or the weight of the neural connection. Overall, high grasping power and adequate simplification capacity are observed in ELM which makes it more suitable for use in energy harvesters.

The mechanisms of the ELM as the training algorithm for the feed-forward neural network with one invisible surface are depicted in Figure 1.8 (c). ELM can generate the network parameters in a random manner.

ELM is widely used in PENG and TENG because of its quick learning capabilities and generalisation capability. Figure 1.8 (d) depicts a deep neural network algorithm known as a stacked auto-encoder (SAE). It is important to note that coding gives unpredictable changes between the input and invisible layers, whereas decrypting gives changes between the hidden and reconstruction layers. SAE can restructure the input data with acceptable error ranges by encryption and decryption process. Figure 1.8 (g) depicts classic AI methods and structures for PENG and TENG. In the design and analysis phase, the nanogenerator is characterised to obtain input data regarding structural design, material properties, geometric properties and external excitation. PENG and TENG inputs are generated by raw data processing. In particular, AI is utilised as a tool to study triboelectric mechanisms in heating elements. A lot of theoretical models are used in past to analyse triboelectric effects, such as the V–Q–X model for TENG in proximity division separation or skidding mode. But these theoretical models were mainly based on some assumptions such as proximity-sharing structures, linear or skidding structures, etc., to simplify them. It is tough to consider the theoretical location of heating elements in practical applications. Or else it becomes complicated. Using AI to analyse a large number of data and establish complicated relationships between inputs and outputs can greatly expand theoretical models of triboelectric mechanisms.

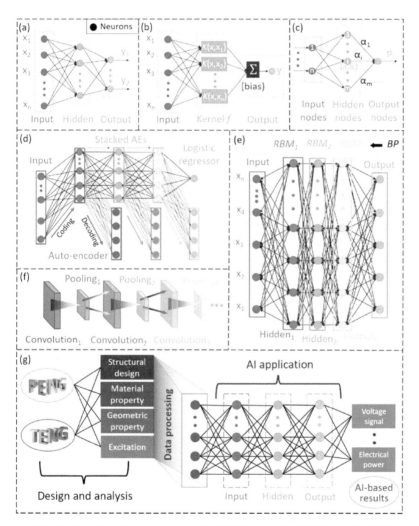

Figure 1.8 Artificial intelligent structure for PENG and TENG [3]. (a) Artificial neural network (ANN). (b) Support vector machine (SVM). (c) Extreme learning machine (ELM). (d) Stacked auto-encoder (SAE). (e) Deep belief network (DBN). (f) Conventional neural network (CNN). (g) Typical procedures and structures of AI-based PENG and TENG.

1.9 Artificial Intelligent in Energy Harvesters

Energy collection and measurement are the two main functions of PENG and TENG, providing a wide scope of use for smart techniques in engineering and life sciences applications. An intelligent system mainly consists of three steps, which include information gathering, information sharing and data analysis. Because of its character as a tool for data analysis, AI is mostly classified as level 3 and has become increasingly important in data processing and analysis. The use of AI in PENG and TENG, in particular, includes (1) big data analysis and sensor processing, (2) nanogenerator design and improvement depending on the requirements and (3) AI algorithm modification and optimisation based on data analysis requirements. At present, AI is largely used to analyse data from sensors, such as grouping and filtering big chunks of information gathered by sensors in PENG or TENG for smart sports devices. Figure 1.9 (a) shows a conceptual iteration in combination with artificial intelligence to forecast the electrical properties of a rotating heating element. The input parameters of the model are the number of components, the rotational speed, the distance between the friction surfaces and the output variables. The AI-enabled prognostication model was constructed to describe the rotary TENG outcomes in the form of kinematics and spatial constraints. Figure 1.9 (b) depicts the TENG-enabled falling juncture allocated numerical scheme and the edge ball judgment system, both of which use AI for data analytics. The wood-based TENG (W-TENG) was suggested as an intelligent table tennis device, and the falling point system is designed by assessing real-time information collection and transforming it into quantitative. Machine learning is used to collect handwriting as input with the help of a TENG tablet and process the information for pattern identification. A multiclass classification model was prepared with the help of an SVM-based flowchart for decision making and the input information was utilised to educate the machine learning model. Figure 1.9 (d) shows a super hydrophobic triboelectric gesture recognition glove. Machine learning has been applied for real-time gesture recognition, and virtual reality (VR) and augmented reality (AR) controls have been implemented. Convolution neural network (CNN) models have been constructed to identify actions and expressions and give appropriate commands via wireless communication. The self-powered HMI shown in Figure 1.9 (e) used TENG for monitoring handwriting in the tapping and sliding modes. CNN was specifically designed with three layers: the input layer, which fed the input data into the neurons; the hidden layer, which included the convolution layer and the linearization

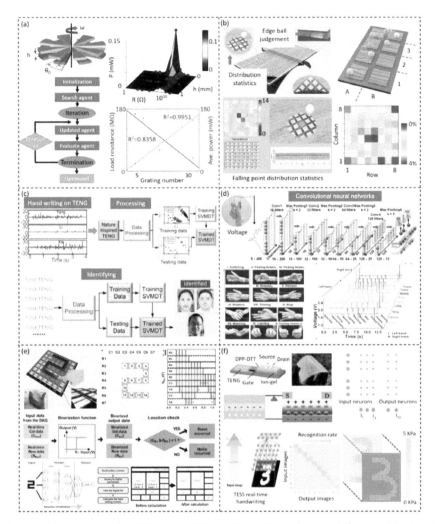

Figure 1.9 Recent AI for PENG and TENG [3]. (a) Theoretical modeling combined with AI to predict the electrical performance of the rotary TENG. (b) TENG-enabled falling point distribution statistical system and edge ball judgment system that use AI for data analytics. (c) Smart self-powered handwriting pad based on the textured TENG. (d) Superhydrophobic triboelectric gloves for gesture recognition. (e) Self-powered HMI that use TENG to monitor handwriting in the tapping and sliding modes. (f) Artificial sensory memory developed based on the TENG matrix.

layer to take out the characteristics of the input and the output layer, which displayed the results.

1.10 Philosophy of AI in Energy Harvesting

AI has received a great deal of research attention in energy harvesting in recent years, primarily because of the incapability of physics-based model mode with the help of first principles. in comparison to the conventional modelling ways that basically use physical principles to learn and forecast a particular response, AI has the ability to understand and manipulate higher-dimensional feature spaces. Depending on the goals and usage, energy-harvesting AI can be classified into technologies like AI algorithms and utilities (i.e., AI-enabled features).

Conventional modelling was mainly based on machine learning algorithms and the new approach includes features such as inference, programming, artificial life, computational development and hindrance satisfaction. A paradigm that applies artificial intelligence with energy harvesting technology. Energy harvesting technology can be divided into three steps, (1) code acceleration apparatus to decrease the economical involvement for the analysis of passive models, (2) creating experimental methods when passive models are not feasible and (3) procuring categorisation equipment. Applications of AI technology in energy harvesting primarily include four layers as depicted in Figure 1.10:

1. the environment layer,
2. hardware layer,
3. software layer and
4. application layer.

At the environment layer, most relations are of low frequency and low amplitude, so artificial intelligence helps control and utilise external activities for energy harvesters. The use of artificial intelligence to adjust and boost the external conditions of this layer to activate piezoelectric and triboelectric materials and generate electricity more effectively. At the hardware level, artificial intelligence helps design and optimise the energy harvester's physical networks and end devices. Hardware is required for the physical network to provide the produced data and to match the inputs and outputs in relation to the structure and materials of the energy harvester. To improve the efficiency and voltage analysis, the generated data must be rectified by signal pre-processing, mining and amplification. The Internet of Things (IoT), or an intelligent cloud can be used for the communication of the data through the wireless gateway, along with this, the data must be converted by another device software like a user interface (UI), man–machine interference, user interference, or modified cloud interface design. At the application

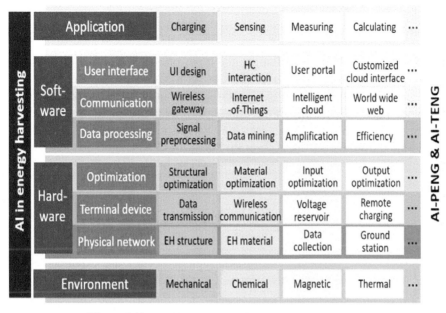

Application		Charging	Sensing	Measuring	Calculating	···
Soft-ware	User interface	UI design	HC interaction	User portal	Customized cloud interface	···
	Communication	Wireless gateway	Internet-of-Things	Intelligent cloud	World wide web	···
	Data processing	Signal preprocessing	Data mining	Amplification	Efficiency	···
Hard-ware	Optimization	Structural optimization	Material optimization	Input optimization	Output optimization	···
	Terminal device	Data transmission	Wireless communication	Voltage reservoir	Remote charging	···
	Physical network	EH structure	EH material	Data collection	Ground station	···
Environment		Mechanical	Chemical	Magnetic	Thermal	···

Figure 1.10 Philosophy of AI for energy harvesting [3].

level, artificial intelligence helps to explore the capabilities of harvesters. In addition to charging with generated electricity, harvesters have also been reported for uses in sensors, measurement, computing and more. These new applications require generators to be AI-optimised and designed to work with other devices.

1.11 Limitation and Future Scope of AI in Energy Harvesting

Artificial intelligence is designed to solve the problems of energy harvesters but, the present AI methods have some limitations. The most important drawback of the AI-based models is that they cannot conclude beyond the limit of the pedagogy of parameter space. AI-based models are generally understood as black box models because most of these methods including machine learning and deep learning cannot elucidate the basic operational correlations between structure, material and power output. Promising power studies of energy harvesters are questionable, even if AI can actually provide superior performance when the quantitative relationship between characteristics and predictions in AI models is unknown. The unavailability of common mechanisms or knowledge of piezoelectric and triboelectric phenomena is also a limitation of artificial intelligence in energy harvesting. AI has proven effective in designing and optimising some types of energy harvesters still it may not be sufficient for other generators as AI models have not been constructed based on common mechanisms. Overfitting or excessive dependency on the aspects and adaptability of the input information are mainly part of AI algorithms and are collaborated energy harvesters creating a new assignment for the problem. Future AI research in PENG and TENG should concentrate on creating more knowledgeable and mighty AI technologies which are able to find solutions to the problems mentioned earlier.

1.12 Conclusion

Fundamental of the energy harvesting and nanogenerators are explained in this chapter along with their applications.

- The future and the growth of energy harvesting lies in artificially intelligent, especially for the nano-generators and its application can be optimised using different algorithms of artificial intelligence.
- The different AI algorithms used in nano-generators and energy harvesters is covered along with the scope of these algorithms for effective energy generation.
- By considering the need for efficient use of energy harvesters and nano-generators for green energy solutions, AI can overcome the flows of the current energy harvesters and nano-generators.
- future research needs to be focused on overcoming the current challenges in these technologies such as mechanisms of current PENG and TENG, its structural design and materials.

References

[1] Sharma, A., Sharma, A., Dasgotra, A., Jately, V., Ram, M., Rajput, S., Averbukh, A.& Azzopardi, B. (2021). Opposition-based tunicate swarm algorithm for parameter optimization of solar cells. *IEEE Access*, 9, 125590-125602.

[2] Sharma, A., Sharma, A., Pandey, J. K., & Ram, M. (2022). *Swarm Intelligence: Foundation, Principles, and Engineering Applications.* CRC Press.

[3] Pengcheng, Jiao, "Emerging artificial intelligence in piezoelectric and triboelectric nanogenerators" Nano Energy 88 (2021) 106227.

[4] Z. L. Wang, J. H. Song, Piezoelectric nanogenerators based on zinc oxide nanowire arrays, Science 312 (5771) (2006) 242–246.

[5] A. Yu, P. Jiang, Z. L. Wang, Nanogenerator as self-powered vibration sensor, Nano Energy 1 (3) (2012) 418–423.

[6] H. Akinaga, H. Fujita, M. Mizuguchi, and T. Mori, Sci. Technol. Adv.Mater. 19, 543 (2018).

[7] S. Y. Chung, S. Kim, J. H. Lee, K. Kim, S. W. Kim, C. Y. Kang, S. J. Yoon, Y. S. Kim, All-solution-processed flexible thin film piezoelectric nanogenerator, Adv. Mater. 24 (2012) 6022–6027.

[8] S. H. Baek, M. S. Rzchowski, V. A. Aksyuk, Giant piezoelectricity in PMN-PT thin Films: Beyond PZT, MRS Bull. 37 (2012) 1022–1029.

[9] P. Jiao, H. Hasni, N. Lajnef, A. H. Alavi, Mechanical metamaterial piezoelectric nanogenerator (MM-PENG): design principle, modeling and performance, Mater. Des. 187 (2020), 108214.

[10] Y. B. Lee, J. K. Han, S. Noothongkaew, S. K. Kim, W. Song, S. Myung, S. S. Lee, J. Lim, S. D. Bu, K. S. An, toward arbitrary direction energy harvesting through flexible piezoelectric nanogenerators using perovskite PbTiO3 nanotube arrays, Adv. Mater. 29 (6) (2017), 1604500.

[11] D. Yao, H. Cui, R. Hensleigh, P. Smith, S. Alford, D. Bernero, S. Bush, K. Mann, H. F. Wu, M. C. Chin-Nieh, G. Youmans, X. Zheng, Achieving the upper bound of piezoelectric response in tunable, wearable 3D printed nanocomposites, Adv. Funct. Mater. 29 (42) (2019) 1–11.

[12] B. Wu, H. Wu, J. Wu, D. Xiao, J. Zhu, S. J. Pennycook, Giant piezoelectricity and high curie temperature in nanostructured alkali niobate lead-free piezoceramics through phase coexistence, J. Am. Chem. Soc. 138 (47) (2016) 15459–15464.

[13] L. Dong, C. Wen, Y. Liu, Z. Xu, Piezoelectric buckled beam array on a pacemaker lead for energy harvesting, Adv. Mater. Technol. 4 (1) (2019), 1800335.

[14] S. K. Ghosh, D. Mandal, High-performance bio-piezoelectric nanogenerator made with fish scale, Appl. Phys. Lett. 109 (2016), 103701.

[15] L. Jiang, Y. Yang, R. Chen, G. Lu, R. Li, L. Di, M. Krailo, K. K. Shung, J. Zhu, Y. Chen, Q. Zhou, Flexible piezoelectric ultrasonic energy

harvester array for bio-implantable wireless generator, Nano Energy 56 (2019) 216–224.

[16] W. S. Jung, M. J. Lee, M. G. Kang, H. G. Moon, S. J. Yoon, S. H. Baek, C. Y. Kang, Powerful curved piezoelectric generator for wearable applications, Nano Energy 13 (2015) 174–181.

[17] X. Niu, W. Jia, S. Qian, J. Zhu, J. Zhang, X. Hou, J. Mu, W. Geng, J. Cho, J. He, X. Chou, High-performance PZT-based stretchable piezoelectric nanogenerator, ACS Sust. Chem. Eng. 7 (2018), 979-958.

[18] A. Mellit, S. A. Kalogirou, Artificial intelligence techniques for photovoltaic applications: a review, Prog. Energy Combust. Sci. 34 (5) (2008) 574–632.

[19] Y. Reich, S. V. Barai, Evaluating machine learning models for engineering problems, Artif. Intell. Eng. 13 (3) (1999) 257–272.

[20] B. Sanchez-Lengeling, A. Aspuru-Guzik, Inverse molecular design using machine learning: generative models for matter engineering, Science 361 (6400) (2018) 360–365.

[21] Y. Reich, Machine learning techniques for civil engineering problems, Computer -Aided Civ. Infrastructure. Eng. 12 (4) (2002) 295–310.

[22] D. M. Dimiduk, E. A. Holm, S. R. Niezgoda, Perspectives on the impact of machine learning, deep learning, and artificial intelligence on materials, processes, and structures engineering, Integr. Mater. Manuf. Innov. 7 (2018) 157–172.

[23] S. M. Zahraee, M. K. Assadi, R. Saidur, Application of artificial intelligence methods for hybrid energy system optimization, Renew. Sustain. Energy Rev. 66 (2016) 617–630.

[24] Z. Wang, R. S. Srinivasan, A review of artificial intelligence-based building energy use prediction: contrasting the capabilities of single and ensemble prediction models, Renew. Sustain. Energy Rev. 75 (2017) 796–808.

[25] Y. Zhou, M. Shen, X. Cui, Y. Shao, L. Li, Y. Zhang, Triboelectric nanogenerator based sel-powered sensor for artificial intelligence, Nano Energy 84 (2021), 105887.

2

Application of the ANN Method in Water Energy Harvesting

Surendra Pal Singh[1], Marga Tashale Amanu[1], and Prashasti Ashok[2]

[1]Surveying Engineering Department, Wollega University, Ethiopia
[2]Department of Geology, Institute of Earth Sciences, India
E-mail: Surendra.geomatics@gmail.com; margatashale@gmail.com;
drprashastisaxena@gmail.com

Abstract

This chapter focuses on artificial neural networks (ANNs), which are used in the field of hydrological (water) energy harvesting. On the other hand, traditional procedures are time-consuming and difficult to implement due to computational analysis. Use in streamflow, rainfall-runoff modelling, water quality modelling, flood forecasting and groundwater modelling are all examples of artificial intelligence functioning [1]. A thorough insight into the hydrologic method under consideration can aid in the selection of the input vector and the creation of a more effective network. The ANN model, which gives high precision for hydrological difficulties and is a more efficient approach for engineering purposes, is discussed in this chapter. Clean energy and fossil fuel resources are tremendously important to modern humanity. Because fossil fuels are becoming limited, the market is more concerned with the marketing of sustainable energy and the reduction of carbon emissions. Several computer models have been used to study the problem of making green energy systems last for a long time.

Keywords: ANN; Rainfall-Runoff; Water Energy Harvesting; Soft Computing.

2.1 Introduction

Modern civilisation depends enormously on renewable energy and fossil fuel resources. Fossil fuel is diminishing, so the market is very much concerned about renewable energy marketing and cutting carbon emissions. To meet their clients' requests, several of these marketers are now relying on biodegradable and renewable energy supplies. While renewable energy sources such as the sun, water (tidal, surface and underground) and wind have been used to successfully construct energy supply and demand balances, numerous experts are arguing that biodegradable resources should be included in the equation. The present levels of penetration of renewable energy resources in the energy market are due to a variety of variables. First and foremost, these supplies are eco-friendly. Second, they are dispersed over the world in various communities. In some instances, a community may have two or more renewable energy sources. Third, the methods for extracting these resources have improved to the point where they may be purchased at a reasonable price. Despite the many benefits of renewable energy systems, experts continue to argue about their reliability and availability. Various computational approaches have been used to tackle the issue of renewable energy system reliability. At this current time, all the statistical matters are difficult without the use of quite a number of soft-computing methods as per the requirements. And in this modern-day time, the location of the water area includes massive statistical facts primarily based on issues due to human area enlargement on the globe and must be addressed in a brief period. In any other case, an imbalance in supply and demand of water sources can be created soon. As human existence is without delay based on water sources for a range of things to do and these matters are to be balanced and desirable if mankind must exist, that is, why is a well-known slogan 'conserve water, conserve life'. Their imbalance can create a major artificial disaster for mankind itself. So, research in this area needs to be updated all the time to keep up with their balance, and soft-computing methods are a very helpful way to deal with the complexity of the water aid problems of today.

2.2 Soft Computing

Soft computing is an emerging type of computing that taps into the remarkable capacity of the human mind to reason and explore in an unpredictable and uncertain environment. Soft computing is generally centred on organically induced approaches such as genetics, development, behaviour, particle warmth, the human nervous system (HNS) and so forth. Soft computing appears to be the only option when we don't have any computational models of challenges (i.e., algorithms), we need to solve a challenging problem in real-time, adapt to changing circumstances and use parallel computing. It serves a wide range of utilitarian applications, including scientific diagnosis, digital devices, computer vision, weather prediction, social optimisation, forecasting, LSI design and sample identification, among others. The ANN technique has been defined under the soft-computing method in this chapter and has been utilised to model complicated hydrological procedures like rainfall-runoff, overland flow, etc. [2] validated one of the most promising hydrology methods [2]. The phrase 'tender computing' used to be first defined by [3], who is recognised as the overview of fuzzy logic. It is a collection of methods relying on laptops and software-based systems as alternatives to difficult computing strategies and that offer realistic information in the same way that human beings deal with them. Soft computing is based totally on realistic input, knowledge about the relationship between entering and output, frequent sense and reasoning, as well as herbal ideas. The typical classification of soft-computing methods used around the world is shown in Figure 2.1. Artificial neural networks (ANNs), support vector machines, fuzzy logic and genetic algorithm are used to address the primary insurance of soft computing in the water quarter. All of these disciplines can be applied to the problem individually or as a combination of methods [2, 3].

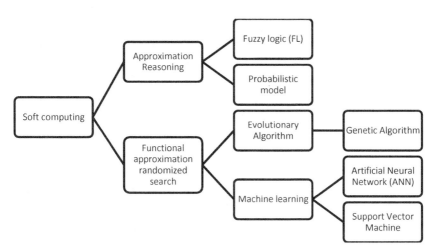

Figure 2.1 Classification of soft computing.

All the soft-computing techniques for purposes are equally necessary for quite a few fields of water resources, such as modelling and simulation of rainfall-runoff, floods, waterlogging, soil erosion, drought salty groundwater, region-wise water desk, ctc., and these are some of the important issues of water administration for agriculture and different needs. As a result, it is preferable to learn a little about soft-computing residences as well as every approach of soft-computing in an instructive and informative manner, and then their direct utility in the subject of the water sector.

2.2.1 Soft-computing properties

One of the foremost purposes of soft computing is to emulate the reasoning and wondering behaviour of human thought as intently as possible, with the aid of which its principal purpose is to comply with the way human thinking works. Further, a precise property of soft computing which makes it fascinating for statistical evaluation is getting to know from experimental facts and deriving their inside relations from interpolating or approximating to produce outputs from formerly unseen inputs by using the usage of outputs from formerly discovered inputs. So, soft computing is primarily based on approximate models as a substitute for precise models, as in the case of challenging computing. Based on the above properties, the real-world troubles, which are pervasively imprecise and unsure, can be regarded as underneath the premises of soft-computing, as the precept of soft-computing is to take advantage of approximation, tolerance for imprecision and uncertainty to gain robustness and tractability with low answer cost. Out of all the categorised strategies under soft computing, the subject of probabilistic reasoning is additionally now and again protected due to its higher management of randomness and uncertainty, although it is not yet been explored a great deal for its utility in the water sector. So with this quick introduction of gentle computing, in addition, we can focus on the person's soft-computing strategies in the subsequent section.

2.3 Artificial Neural Networks (ANN)

In a sense, human brains can explain real-world conditions in a way that computer systems cannot. In the 1950s, neural networks were built for the first time to resolve this issue. An artificial neural network is an attempt to emulate a community of neurons that make a human talent so that computer systems can be able to analyse matters and humanly make decisions. ANN is made with regular computer programming as if the lines of code were linked together like talent cells. The natural neural community, which consists of a billion interconnected neurons in the brain, has been used to excite artificial neural networks. Thanks to developments in the field of statistics processing, artificial neural networks were used to represent the talent's massively parallel processing and scattered storage houses. An ANN is a mathematical shape that can express unpredictable, sophisticated and nonlinear systems by correlating their input and output. An artificial neural network (ANN) is a data-processing computer that consists of a highly interconnected network of simple processing components known as neurons. Within the community, the neurons of an ANN are arranged into layers, with the neurons of one layer connected to those of the next layer. In an organic neural network, the weight is equal to the electricity of the indicators, which is the electricity of these connections between two neighbouring layers. During the studying or training process, the weights of the connections are changed until the inputs lead to the desired output.

Integrated learning procedures are mandatory for weight tuning to achieve the desired result based on the training data provided to the community. The multi-layer perceptron (MLP) is one of the most familiar neural network types. An MLP is a community made up of three distinct types of layers: input, hidden and output layers [3]. A feed-forward neural network is referred to as an MLP. Because all the statistical records move in one direction, it is known as feed-forward. The use of connecting the neurons with the next neurons prevents feedback. A whole connected MLP with a hidden layer is shown in Figure 2.2. Feed-forward back-propagation is a supervised training system.

While using a supervised coaching approach, the community should be given both pattern inputs and outputs. The back-propagation schooling method forecasts a record's output and compares it to the result If there is an

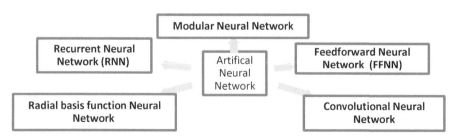

Figure 2.2 Types of artificial neural networks (ANN).

error, the weights of several layers are changed by moving backwards from the output layer to the enter range [4]. Research has proven that ANN models are beneficial for machining and getting to know equipment for electricity structure analysis. In an attempt to use this equipment to analyse strong struc-ture parameters, pupils have succeeded in figuring out specific parameters for special purposes [5–8]. They have also made significant contributions to the overall performance of ANN models. One of their accomplishments is designing computational algorithms that combine international optimisation techniques with neural network modelling. These algorithms have been used to solve classification and prediction issues successfully. Reynolds et al. [9], for example, discussed the use of a combined ANN and genetic algorithm model for strength management analysis. Artificial neural networks provide a convenient way to address complicated and poorly defined problems. They can analyse from illustrations, are fault-tolerant in the sense that they can cope with noisy and incomplete data, can handle nonlinear problems and, as soon as educated, can manage predictions and generalisations at excessive velocity. Artificial neural networks have transformed the way educational fact sets are learned, computed, analysed and outputted.

2.4 Application of ANN in Hydrology

2.4.1 In stream-flow modelling

Modelling streamflow allows for the more methodical operation of water sources below commercial, administrative, scientific and political agendas. Streamflow is widely used to estimate watershed runoff and should be considered a phase of the prior section. The emphasis here is on articles that have dealt with streamflow directly, except for papers that have used precipitation as an input. Streamflow prediction was once considered an inter-mediate objective in several studies. A variety of three-layered ANN designs were studied. These first findings reveal that ANNs are an effective tool for forecasting streamflow. In, [10], the authors make use of an artificial neural network method along with the cascade-correlation algorithm to estimate the Huron river's drift forecast.

The results show that artificial neural network models (ANN) are effective in making healthy changes in the flow history. The authors in [11] employed a back-propagation method to calculate month-to-month streamflow. In another study [12], the authors projected streamflow using the recurrent neural network (RNN) model. Scholars examined the utility of artificial neural networks (ANNs) in predicting implied month-to-month streamflow and in contrast with auto-regressive (AR) models. Hybrid ANN models for forecasting the day-by-day streamflow have been implemented. Three types of hybrid artificial neural networks (ANNs) were used: threshold-ANN (T-ANN), cluster-ANN (CANN) and periodic-ANN (P-ANN).

2.4.2 In water quality modelling

The flow rate, pollutant load, mode of conveyance, preliminary conditions, water stages and several site-specific characteristics all have an impact on water quality. For such intricate and nonlinear issues, the application of ANN is ideal. Maier and Dandy [13] tested the efficiency of ANNs for measuring salinity at the Murray Bridge on the Murray River in South Australia. Rogers and Dowla [14] used an ANN that was trained using a solute transport model to perform optimisation experiments in ground-water treatment. The returned propagation coaching algorithm was used to train a multilayer feed-forward ANN. The results obtained by way of this technique have been consistent with those ensuing from a traditional optimisation method, the use of the solute transport model and nonlinear programming, the usage of a quasi-studies was performed on the utility of ANN for water satisfactory management. Earlier studies yielded a comprehensive conclusion about the use of the ANN model for predicting contaminants in floor water. A neural network model for anticipating the presence of a one-of-a-kind harmful metal in groundwater was developed as part of their research. An ANN model was once used to forecast wastewater volumes in the future.

2.4.3 In groundwater modelling

Rizzo and Dougherty [15] used a neural kriging algorithm to characterise aquifer characteristics. A three-layer neural network with the counter propagation approach was paired with kriging to calculate hydraulic conductivity. The entry nodes indicated the locations of commenting points. The output nodes calculate the type of hydraulic conductivity at a range of places. They found that ANNs could be a beneficial tool in geohydrology when used to precisely classify aquifers. As inputs, they used daily rainfall, past water table locations and potential evapotranspiration as inputs. For the first and only time, the output was the present area of the water table. After coaching the employment of located values, they discovered that a feedforward ANN with three layers might be able to estimate water table elevations satisfactorily.

For groundwater degree predictions in a shallow aquifer, Nayak et al. [16] employed an artificial neural network (ANN) technique. The findings suggest that ANN models are capable of forecasting water tiers up to four months in advance reasonably accurately. Nourani et al. [17] estimate groundwater level (GWL) in Ardabil, northwest Iran, using a mathematically based comprehensive model.

2.5 Case Studies: Application of ANN in Water Resources

There are models which are developed by setting up the linear relationship between input and output by fending off the complicated bodily laws. With the help of a unit hydrograph, such a linear rainfall-runoff relationship is clearly shown. However, these models do not account for the nonlinear dynamics of the rainfall-runoff transition. The use of the ANN approach in rainfall-runoff modelling is a cutting-edge advance in the system-conceptual modelling strategy. The advantage of the ANN method is that it does not necessitate specific expertise in catchment features; it truly develops a correlation between the rainfall (input) and runoff (output) on the foundation of understanding via the neural network's education system. Thus, the bodily traits, even though now not given weightage one at a time, are very much an inherent part of the model.

2.5.1 Rainfall-runoff (RR) process

The fundamental grasp of RR together with the conversion of rainfall into runoff is imperative if one is to create a RR model. This segment offers quick important points about the processes and changing aspects of the complicated RR method. The hydrological cycle displaying the RR event is introduced in Figure 2.3. Oceans are commonly used to determine the amount of water that can be reached on the earth's surface. Once it begins to evaporate, it goes to the surroundings, ensuing in specific conditions, vapour condensate from clouds turns out to be rainfall. Rainfall falls at once on both the ocean and rivers, which transport it to the ocean. Different probabilities exist for the waterfalls on the earth's floor, which can be intercepted by way of vegetation and evaporate, additionally infiltrating the soil. The infiltration technique brings in unsaturated factors absorbed via vegetation, in which the water returns to the environment via transpiration processes. The saturated region of water accountable for basin runoff is a section of subsurface water. The phenomenon of subsurface water flowing again into the watercourses is

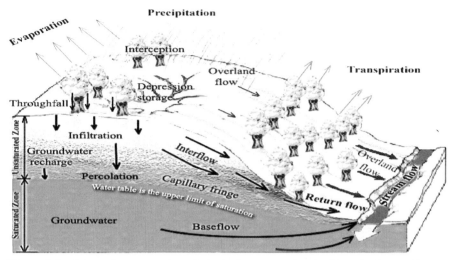

Figure 2.3 Schematic presentation of physical processes involved in rainfall-runoff modelling (David G. Torton, 2003).

regarded as seepage. To exhibit the rainfall and runoff versions, a hydrograph [18]. The version of rainfall match introduced by way of hydrograph consists of three stages: (a) rising limb, (b) falling limb and (c) recession limb.

As per Chow, Maidment and Mays (1988), there are three extraordinary kinds of overflow, that is, surface, subsurface and groundwater. This phase describes several kinds of overland and different flows, as proven in Figure 2.5.

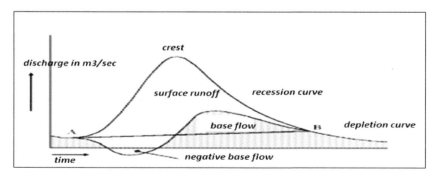

Figure 2.4 Hydrograph showing rainfall-runoff process.

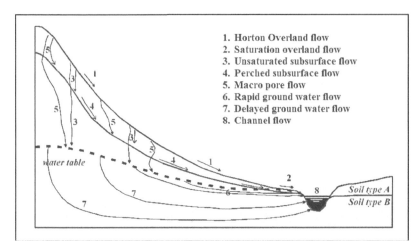

Figure 2.5   Various surface flow processes (Rientjes and Boekelman, 2001).

2.5.2 Surface runoff

The portion of the runoff that flows over the floor and through channels to reach the catchment outlet is known as surface runoff. The following is a list of drift strategies that have been converted to floor runoff. Overland drift is the movement of water over land using a sheet flow (i.e. thin water layer) or as converging flow into rill flow (i.e. small rills). Overland flow can be divided into two types: (1) Horton flow is generated by the infiltration additional mechanism. (2) The saturation extra mechanism is used to cause saturation overland flow. This flow is caused by the soil becoming saturated as a result of the water desk's upward push to the ground surface [19]. (3) Water infiltration in the subsurface is used to create unsaturated subsurface flow. (4) Due to hydraulic head gradients, perched subsurface drift occurs in saturated subsurface circumstances when water flows in lateral directions and waft is difficult to manage. (5) Rapid groundwater glide is the segment of groundwater glide that is released at the upper section of the first unsaturated subsurface domain. (6) Delayed groundwater flows with the flow and is discharged as groundwater in a reduced amount.

2.5.3 Rainfall-runoff (RR) modelling approaches

The RR models establish the relationship between rainfall and runoff for a basin. The method of rainfall into the runoff is used to forecast the stream-flow beneficial for water planning, which consists of infrastructural works, excessive flood and drought variant control, water generation, irrigation and drainage structure design. Because of the spatial-temporal variability of basin facets and rainfall patterns, as well as the large number of variables involved in the modelling of physiological tactics, rainfall-runoff conversion is one of the most challenging hydrologic phenomena [20]. The following is the classification of the R-R models.

2.5.4 Physically based RR models

This model consists of an answer to partial differential equations, which indicates an excellent illustration of flow tactics over the catchment (Figure 2.6). Partially differential equations are frequently solved by discretising space-time dimensions into a discrete set of nodes [21]. If a model represents a system with specific zones of space (i.e. the system is subdivided into spatial units of equivalent or non-equal sizes), it is called a distributed model. Physically based models often use two-dimensional, but sometimes three-dimensional, distributed data [22].

2.5.5 Conceptual RR models

A model can be considered a conceptual model when the physics of a device is simulated using mathematical relations and the model factors have no more than a passing similarity to the actual ones [21]. This category includes models that can't be grouped as extraordinary pragmatic models. The basic

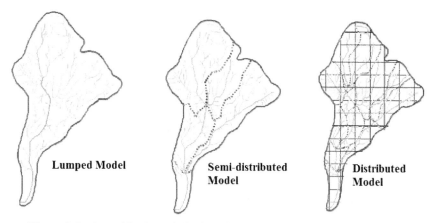

Figure 2.6 A combined, semi-distributed and distributed model approach [22].

idea behind conceptual models is that flow is linked to storage using the conservation of mass equation and a transformation model.

2.5.6 Empirical RR models

RR modelling can be done in a solely analytical context based on data from a basin's inputs and outputs. The basin is treated as if it were a black box, with no mention of the internal operations that monitor the transformation of rainfall into runoff [23]. This kind of model is generally utilised when family members emerge as complicated and challenging to describe. Empirical models are frequently utilised at the catchment scale, where only small facts about the hydrologic system are reachable. The multiple linear regression (MLR) model is a well-known example of empirical R–R modelling. Black box models include artificial neural networks (ANNs). The distinction between these two types of black-box models is that time-series models predict only the most recent values of the variable, whereas ordinary black-box models predict the entire time series. As a result, time series models are referred to as local models, in contrast to other black box models' global approaches. Because ANNs are black-box models, they can also be used to model time-series data (e.g. both model input and model output are based on catchment output).

2.6 ANN Application in Water Energy Harvesting

French et al. [24] used ANN-based modelling for the first time in the field of hydrology. Following that, ANN's mission has always been in hydrology. For hydrograph prediction, Halff et al. [25] employed a three-feed-forward ANN. The ANN model was proposed by Raman and Sunil Kumar [26] for estimating month-to-month rainfall. ANNs update their structure primarily based on external and interior records that feed to the device community during the getting-to-know phase. This mimics the working of the human talent network, which learns from rides barring any previous facts about the machine. For the administration of water, sources require planning, operation and improvement to pick out the issues that are associated with the quality and extent of water. With the increase of pressures on water assets due to anthropogenic and local weather change, the approach to predicting precisely excessive scenarios in water administration is wanted by decision-makers and watershed managers. Soft computing is an automatic wise predictor to predict complicated water area troubles by imparting great results, which is challenging to clear up by using challenging computing [27]. Soft computing offers approximate outcomes with a low degree of estimation and is quicker in contrast to challenging models. In the past three decades, more than a few soft computing strategies have been developed, which are the integration of natural constructions and computing methods, that is, some of ANN, genetic algorithms, fuzzy logic and fusion techniques [28].

2.7 Rainfall-runoff (RR) Modelling using ANNs

ANNs indicate the most accurate equipment for hydrological modelling, that is, rainfall-runoff events, from previous decades. ANNs have been tested to supply first-class results in comparison to different techniques. ANNs are in a position to map the underlying relations between input and output variables besides the short facts of the procedures' underestimation by using the ideal set of community parameters via the education phase. The area introduced the dynamics of rainfall-runoff conversion below a catchment, one of a kind of occasion associated with this system. The ANNs are used to model rainfall-runoff events to understand several facts about distinctive troubles associated with the sketch of ANN models for RR modelling. ANN has proven to be one of the best techniques to model complicated hydrological methods like the rainfall-runoff manner and different hydrological events. In most of the research, ANNs have confirmed higher consequences in contrast to different techniques. Aside from a prior understanding of the technique under research, ANNs can map the link between input and output data [12]. Several models, that is, artificial neural networks, bodily-based, black container and conceptual models, have been applied to simulate complicated hydrological strategies such as the rainfall-runoff process. Overland float and other hydrological events [30, 31]. It has been established to become one of the most effective pieces of hydrological equipment [32]. Due to its complexity and spatio-temporal variation, only a few models can precisely simulate this extraordinarily nonlinear method to forecast upcoming river discharge, which is required for protected and low-cost factors in hydrologic and hydraulic engineering diagrams and water administration drives.

Kumar et al. [2] utilised ANN methods to consider the rainfall-runoff match of the Sarada River Basin with backpropagation algorithm, calibration and validation with their competencies, which influence the input dimensions of the rainfall-runoff model.

The observed and computed daily runoff using the ANN model during calibration (2001–2007) and validation (2008–2010) are shown in Figure 2.9 (a) and (b). The coefficient of determination value was found to be 0.81 during calibration and 0.78 during validation. Figure 2.10 depicts a comparison of experimental and calculated runoff during the calibration and validation periods for the years 2008, 2009 and 2010, demonstrating that the predicted daily runoff values match well with the observed runoff values during validation. The excellent correlation between the measured and simulated runoff values throughout the calibration period is evidenced by the high coefficient of determination values of 0.81 and model efficiency of 78.37%. To recognise the link established via the ANN model, Kalteh [32] used the ANN model to anticipate rainfall-runoff tactics and defined unique processes such as the neural interpretation diagram, Garson's algorithm and randomisation method.

The findings show that ANNs are an excellent tool not only for excellent modelling but also for displaying perception from the realised relationship, which assists the modeller in explaining the procedure under investigation. Ozesmi and Ozesmi [22] established the neural interpretation plan (NID) to interpret the joining weights of an educated ANN model based entirely on the imagining of their magnitude and direction. The rainfall-runoff approach,

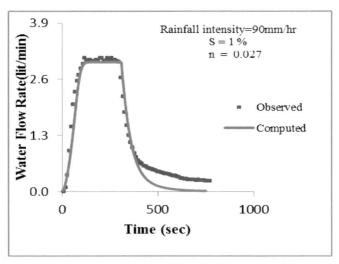

Figure 2.7 Comparison of observed and computed hydrograph for rainfall intensity 90 mm/hr at 1% slope of the plane.

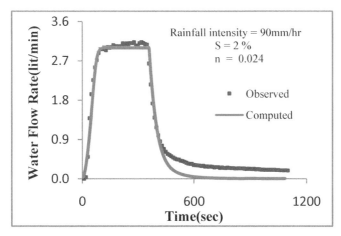

Figure 2.8 Comparison of observed and computed hydrograph for rainfall intensity 90 mm/hr at 2% slope of the plane.

in which the amplitude and route of the correlation are characterised as line thickness and country, was once identified using NID (dotted or solid). The magnitude's thickness has been scaled to the fee of the magnitude's weight; strong connection weights have been used to demonstrate massive connections, while dotted connection weights have been used to show weak connections. The weights for the input-hidden layer connection and the weights for the output-hidden layer connection are represented by two types of connection weights. As a result, the rainfall-runoff relationship, or the

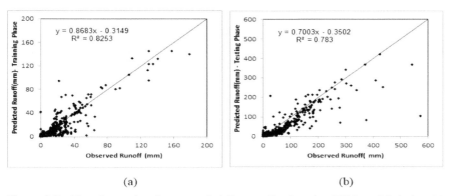

(a)   (b)

Figure 2.9 The observed and computed daily runoff using the ANN model during (a) calibration (2001–2007) and (b) validation (2008–2010) [2].

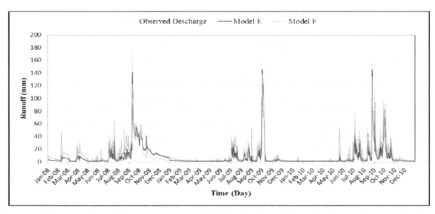

Figure 2.10   Examined and computed runoff for the duration of the calibration and validation period.

relationship between entering and output variables, is determined using the input-hidden and hidden-output layer connection weights. Garson [33] proposed a method for determining the relative influence of each entry variable in the system output by using connection weights obtained from ANNs.

Three-layer feed-forward MLP models have been utilised to estimate the month-to-month runoff in a basin in Iran. Nine input variables have been utilised for the feed-ahead MLP model and one output variable. Figures 2.3 and 2.4 exhibit the scatter diagrams of the found and computed month-to-month runoff located by the MLP model for the coaching and validation period. This suggests runoff used to be nicely simulated with the developed MLP model. Figure 2.13 indicates the NID for the ANN rainfall-runoff model to depict the influence of every entered variable in opposition to hidden neurons on the output variable. To decide quantitatively the contribution of all entered statistics on the output, Garson's algorithm has been utilised as proven in Figure 2.14, which represents the consequences of an entering variable from 4.12% to 19.73%. The effects demonstrate that the ANN model predicts first-class outcomes in contrast to every other model.

Antar et al. [34] utilised the ANN model to predict the rainfall-runoff match in the Blue Nile basin. The effects of comparative evaluation with a physical model show that the ANN approach has proven to be quite effective in modelling rainfall-runoff processes.

Input variables have been taken following the complete variables. The number of hidden nodes has been set equal to the variety of entering nodes. The effects of sensory evaluation of several iterations are a thousand the assessment of discovered and computed each day hydrographs at the Diem station ensuing from the calibration scan, which represents the model may want to simulate most of the flood peaks, is proven in Figure 2.11. The ANN model should reproduce the peaks of the floods successfully, as given in Figure 2.12. The kinematic wave principle is used to route the drift inside a community system. The regular parameters have dynamic parameters referred to as nation variables, which explain the fame of soil moisture deficit on an everyday basis. The everyday operation of that dispensed rainfall-runoff machine on one specific day makes use of the modern-day rainfall values of that day and the country's variables of the day earlier than to replace the country variables of that precise day. The simulated allotted rainfall-runoff system over 1997–1999 was carried out and validated with ANN model results.

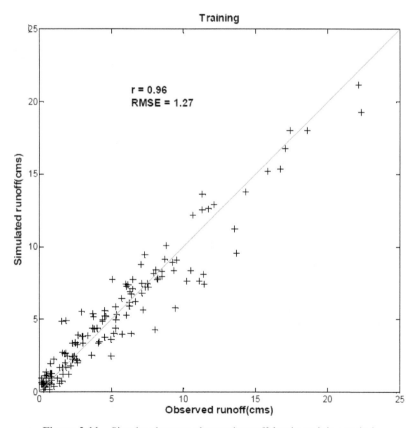

Figure 2.11   Simulated versus observed runoff for the training period.

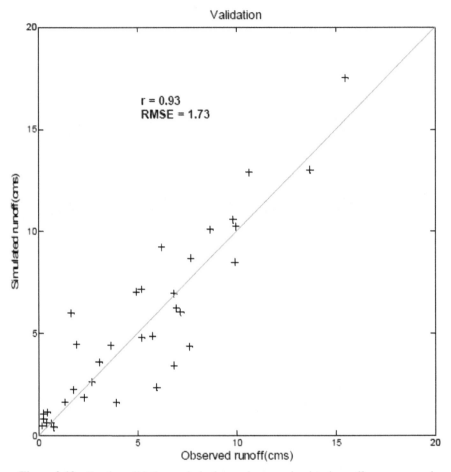

Figure 2.12   For the validation period, observed versus simulated runoff was compared.

Finally, as compared to other models, ANNs compute rainfall-runoff events better. The results show that the ANN model is applied and, if trained with suitable sets of input, has great potential to compute the rainfall-runoff process. The ANN model does not require an intensive data set as compared to distributed physically-based models. However, the model's adherence to the observed data reveals an underestimating of the highest streamflow values. Peak discharges are linked to surface runoff, which occurs because of heavy rainfall. However, the size of the maximum streamflow is affected not only by the intensity of the rainfall but also by other parameters such as soil moisture, terrain and land use, making predicting maximum occurrences difficult in continuous simulations. Otherwise, there was a stronger agreement between the curves during recessions. The fact that groundwater runoff is more common than direct surface runoff can explain why the models work better during recessions. Groundwater runoff is easier to model than direct surface runoff because it comes from the discharge of the aquifer, according to Darcy's law for fluidity in porous media (Figure 2.15).

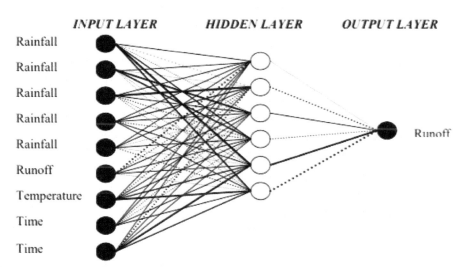

Figure 2.13 Three-layer artificial neural network interpretation diagram (AN-NID) event (Kalteh, 2007)30.

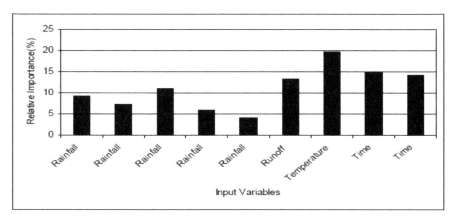

Figure 2.14 Garson's approach determines the comparative importance of each input variable (%) in simulating runoff.

Rajurkar et al. [31] applied the ANN technique to estimate daily flows in the Narmada River in Madhya Pradesh, India, during monsoon flood episodes. A linear multiple-input single-output (MISO) model combined with artificial neural networks (ANN) gives a better approximation of the rainfall-runoff process in large basins than linear and nonlinear MISO models.

Figure 2.16 (a)–(c) shows the plot nicely unfolded over the best line for three enter conditions, whereas for one and two enter conditions, the plot was once moved on aspect due to the division of the large basin into a small one. The related scatter plots for the linear MISO model are displayed in Figure 2.16 (d) reveal a considerable scatter and shift from the ideal line, indicating that the error values have persistence and time variance. The nonlinear MISO model, as shown in Figure 2.16 (e), has a lower plot than the linear MISO model.

Figure 2.16 (d) depicts the plots between measured runoff hydrographs calculated using the ANN-MISO model for validation, and it was discovered that for the duration of validation duration using single rainfall enter the usage of ANN approach, expected time to top was once poorly estimated. As shown in Figure 2.16, the outcomes of the three entrance circumstances for ANN-MISO, linear MISO and nonlinear MISO approaches for validation periods are graphically in contrast to the measured runoff. The hydrograph peak was consistently underestimated using the linear MISO model. The peak discharge anticipated by both models, on the other hand, was better.

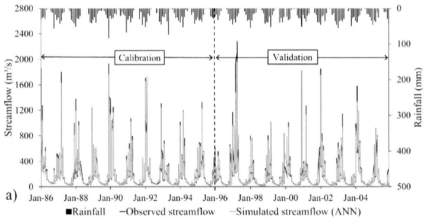

Figure 2.15   Observed and simulated hydrographs using ANN during the calibration and validation periods (Rodrigues et al., 2021).

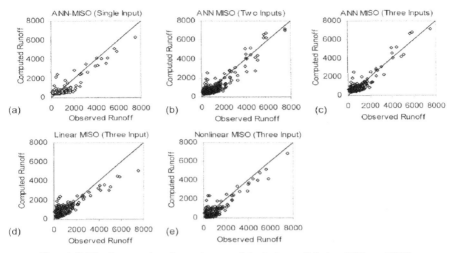

Figure 2.16 Scatter plots for various models during validation (Aitken, 1973).

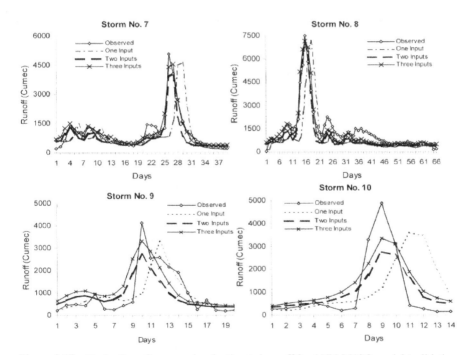

Figure 2.17 Evaluation of measured and estimated runoff for ANN-M1SO model (validation period) [31].

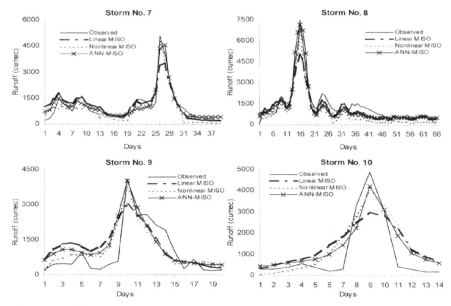

Figure 2.18   Difference of measured and estimated runoff by different models for three inputs. A condition during validation [31].

2.8 Implementation of ANN-RR (Rainfall-Runoff) Models

Figure 2.21 delivers a conceptual framework of those stages the neuro hydrologist must perform when emerging ANN rainfall-runoff models (RR-models).

Step 1. Data collection:
To make sure that all the data are available for research for quantity and quality [20].

Step 2. Selection of predictand:
Clearly express the created model application, keeping in mind that if the data is highly variable, it may be more accurate to model changes influx. Check to see if the data is appropriate for such a model.

Step 3. Data preprocessing (Stage1):

3.1 Data cleansing:
Remove any major rising or plunging trends from the equation. Eliminate temporal components and/or filter the data as necessary to reduce noise and highlight the dominant signal.

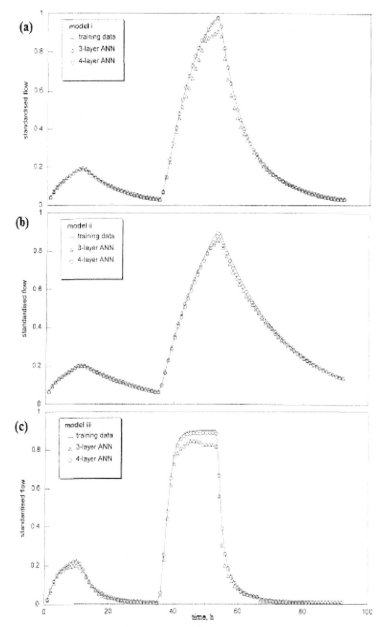

Figure 2.19 The input and output data from three conceptual catchment models were utilised to train three- and four-layer ANNs: (a) model I; (b) model II and (c) model III. Only two incidents have been chosen for illustration clarity.

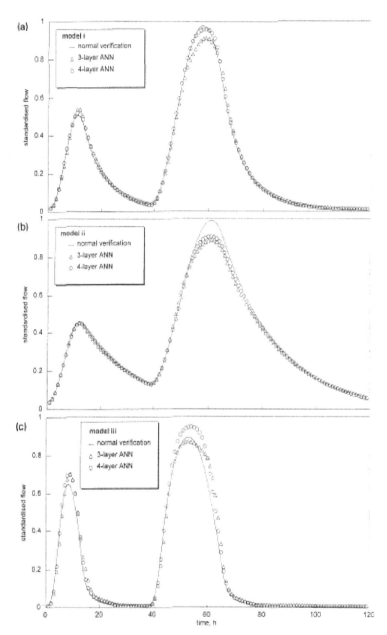

Figure 2.20 Validation of three- and four-layer ANNs trained on input and output data from three conceptual catchment models: (a) model I; (b) model II and (c) model III. For the sake of illustration, only two events have been chosen.

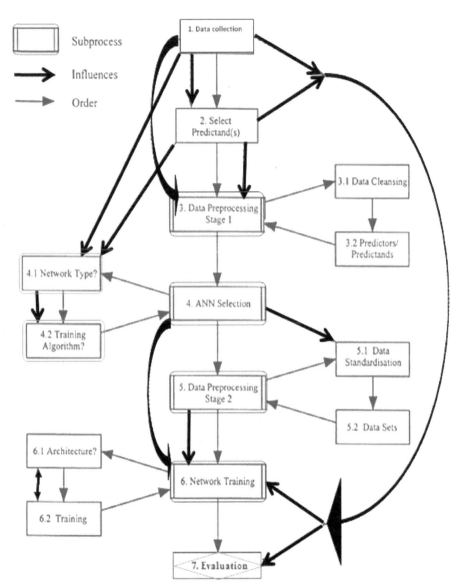

Figure 2.21 ANN model for rainfall-runoff process.

3.2 Predictors/predictands:

Determine the most important predictors for the chosen predictand. If necessary, use main components to reduce the number of predictors.

Calculate appropriate moving averages for each predictor and determine acceptable lag durations for each predictor. Neural networks, ARMA models and autocorrelation functions can all be used to accomplish this. Identify any other processes, such as storm sequencing, that may be significant.

Step 4. ANN selections:

4.1 Network type:
Choose the best network type for your needs. While there is no one-size-fits-all solution, both MLP and RBF neural networks are good places to start when it comes to prediction challenges. The logistic sigmoid or hyperbolic tangent functions are widely utilised as a starting point for MLPs. RBFs using Gaussian basis functions are the most common.

4.2 Training algorithm:
To adjust weights and biases, as well as design network architecture, choose a suitable training procedure. Within the range of 0.01–0.9, choose suitable values for learning constraints (momentum and learning rate).

Step 5. Data preprocessing (stage 2):

5.1 Data standardization: Standardize data to the ranges [0,1], [−1,1], [0.1, 0.9], etc., or normalize the data, depending on the algorithm used.

5.2 Data sets: Divide the data into appropriate calibration, analysis, and validation sets to create cross-validation data sets. Cross-training, on the other hand, should be employed for smaller data sets.

Step 6. Network training:

6.1 Architecture: The number of hidden layers and the number of nodes in these layers must be specified.

6.2 Training: Train several networks using the adjustment and test data.

Step 7. Evaluation:
Select error measures that apply to the model output and objective. Evaluate results with those obtained from alternative model configurations.

2.9 Conclusion

Finally, the connection weights of trained ANN models can be used to construct hydrological 'rules' via sensitivity analysis or rule extraction approaches [24]. As a result, ANNs may be able to uncover previously unseen relationships within hydrological 'black boxes'. The ANN model was used by Senthil Kumar et al. [12] to estimate rainfall-runoff events in two Indian river basins. The study examined the effectiveness of MLP and RBF-type neural network models for rainfall-runoff modelling in detail. Both the MLP and RBF network models were carefully looked at in terms of how well they generalised, how well they predicted the flow of water and how much uncertainty there was in the predictions.

The findings of this study reveal that the type of network used has a considerable impact on model prediction accuracy and that both networks have advantages and disadvantages. For example, the right number of hidden nodes in an MLP must be found through a long process of trial and error, while the structure of an RBF can be found through a good training method.

References

[1] Rakesh T., Tanweer, Desmukh S., "Application of Artificial Neural Network in Hydrology- A Review", International Journal of Engineering Research & Technology (IJERT) ISSN: 2278-0181 IJERTV4IS060247 www.ijert.org (This work is licensed under a Creative Commons Attribution 4.0 International License.) Vol. 4 Issue 06, June-2015.

[2] Kumar P. S., Praveen T. V. & Prasad M. A. "Artificial Neural Network Model for Rainfall-Runoff" International Journal of Hybrid Information Technology, Vol.9, No.3 (2016), pp. 263-272, 2016.

[3] Zadeh L. A., "Fuzzy Logic, Neural Networks and Soft Computing," Communication of the ACM, Vol. 37, Issue.3, pp. 77-84, 1994.

[4] Sivanandam S. N., Deepa S. N., "Principles of Soft Computing", Second Edition, Published by Wiley India, New Delhi, 2011.

[5] Rosenblatt R., "Principle of Neurodynamics", Spartan Books, New York, pp 23-35, 1962.

[6] Minsky M. and Papert S., "Perceptrons: An Introduction to Computational Geometry", MIT Press, Cambridge, Mass., pp 43-80, 1996.

[7] Jain A. K. and Mao J., "Neural Networks and Pattern Recognition in Computational Intelligence: Imitating Life", IEEE Press, Piscataway, pp.194-212, 1994.

[8] Haykin S., "Neural Networks: A Comprehensive Foundation", MacMillan College Publishing Co., New-York, pp.115-133, 1994.

[9] S. Rajput, E. Bender, M. Averbukh, Simplified algorithm for assessment equivalent circuit parameters of induction motors, IET Electric Power Applications, 14 (2020) 426-432.Karunani.thi N., Grenney W. J., Whitley D., and Bovee K., "Neural networks for river flow prediction", J. Comp. in Civ. Engrg., ASCE, 8(2), 201–220, 1994.

[10] Markus M., Salas J. D., and Shin H.-K., "Predicting stream flows based on neural networks", Proc., 1st Int. Conf. on Water Resour. Engrg., ASCE, New York, 1641–1646, 1995.

[11] Senthil Kumar A. R., Sudheer P. K., Jain S. K., and Agarwal P. K., "Rainfall-runoff modelling using artificial neural networks: comparison of network types", Hydrological Processes, John Wiley & Sons, Vol. 19, PP 1277–1291, 2005.

[12] Maier H. R., and Dandy G. C., "The use of artificial neural networks for the prediction of water quality parameters", Water Resources Res., 32(4), 1013, 1996.

[13] S. Rajput, E. Farber, M. Averbukh, Optimal Selection of Asynchronous Motor-Gearhead Couple Fed by VFD for Electrified Vehicle Propulsion, Energies, 14 (2021) 4346.

[14] S. Mengesha, S. Rajput, S. Lineykin, M. Averbukh, The effects of cogging torque reduction in axial flux machines, Micromachines, 12 (2021) 323.Nayak P. C., Sudheer P. K., and Jain S. K.," Rainfall-runoff modelling through the hybrid intelligent system", Volume43, Issue7, First published: 13 July 2007 https://doi.org/10.1029/2006WR004930.

[15] Nourani V., Baghanam A. H., Adamowski J., Kisi O., "Applications of hybrid wavelet–Artificial Intelligence models in hydrology: A review", Journal of Hydrology, Volume 514, 6 June 2014, Pages 358-377, 2014.

[16] Zhang G. P., Fenicia F., Rientjes T. H. M., Reggiani P., Savenije H. H. G., "Modeling runoff generation in the Geer river basin with improved model parameterizations to the REW approach", Volume 30, Issues 4–5, Pages 285-296, 2005, Physics and Chemistry of the Earth, Parts A/B/C

[17] Dunne T., "Relation of field studies and modelling in the prediction of storm runoff", Journal of Hydrology, Volume 65, Issues 1–3, Pages 25-48, August 1983.

[18] Tokar S. A. and Peggy J. A. "Rainfall-runoff modelling using artificial neural networks" Journal of Hydrologic Engineering, ASCE, Vol. 4(3), PP 232-239, 1999.

[19] Govindaraju R. S., "Artificial neural networks in hydrology, II: hydrologic applications. Journal of Hydrologic Engineering, 5, 124-137, 2000.

[20] Ozesmi, S. L. & Ozesmi, U. "An artificial neural network approach to spatial habitat modelling with interspecific interaction". Ecol. Modelling. 116, 15-31, 1999.

[21] Beven K. J., "Rainfall-runoff modelling: The primer (2nd ed.): Wiley-Blackwell, 2012.

[22] French M. N., Krajewski W. F., and Cuykendal R. R., "Rainfall forecasting in space and time using a neural network. "J. Hydrol., Amsterdam, 137, 1–37, 1992.

[23] Sharma, A., Sharma, A., Jately, V., Averbukh, M., Rajput, S., & Azzopardi, B. (2022). A Novel TSA-PSO Based Hybrid Algorithm for GMPP Tracking under Partial Shading Conditions. Energies, 15(9), 3164.Raman H., and Sunilkumar N., "multi-variate modelling of water resources time series using artificial neural networks" Hydrological Sci., 40, 1995.

[24] Smakhtin V. U.,"Low flow hydrology: a review", Journal of Hydrology, 147–186, www.elsevier.com/locate/jhydrol, 2001.

[25] Nayak P. C. and Sudheer K. P., "Fuzzy model identification based on cluster estimation for reservoir inflow forecasting", Hydrological Processes, 827–841 Published online 11 June 2007, Wiley InterScience (www.interscience.wiley.com) DOI: 10.1002/hyp.6644

[26] Kalteh, A. M., "Rainfall-Runoff Modelling Using Artificial Neural Networks (ANNs). Department of Water Resources Engineering, Lund Institute of Technology, Lund University, 2007.

[27] Dawson C. W. and Wilby R. L., "Hydrological modelling using artificial neural networks" Progress in Physical Geograph, Vol. 25(1), PP 80–108, 2001.

[28] Rajurkar M. P, Kothyari U. C. and Chaube U. C., "artificial neural networks for daily rainfall-runoff modelling" hydrological sciences journal, vol. 47(6), pp 865, 2002.

[29] Kalteh A. M., "Rainfall-runoff modelling using artificial neural networks (ANNs): modelling and understanding" Caspian Journal of Environmental Sci., Vol. 6 No.1 pp. 53-58, 2008.

[30] Garson, G. D., "Interpreting Neural Network Connection Weights. AI Expert, 6, 47-51, 1991.

[31] Antar M. A., Elassiouti I. and Allam M. N., "Rainfall-runoff modelling using artificial neural networks technique: a Blue Nile catchment case study" Hydrological Processes, Wiley International Science Publication, Vol. 20, PP 1201–1216, 2006.
[32] Rodrigues, J. A. S.; Andrade A. C. D. O; Viola M. R.; Ferreira D. D.; Mello C. R. D. M.; Thebaldi M. S., "Hydrological modeling in a basin of the Brazilian Cerrado biome", January 2021, Ambiente & Água - An Interdisciplinary Journal of Applied Science 16(1):18, 2021.

3

Artificial Intelligence (AI) in Electrical Vehicles

Dinesh Kumar Bajaj and Pandi Siddharth

MIT-ADT University, India
E-mail: dineshkumar.bajaj@mituniversity.edu.in;
siddharth.pandi@mituniversity.edu.in

Abstract

For decades, the whole world's energy production has been based mostly on unsustainable fossil fuels such as coal, oil and gas, resulting in serious issues such as climate change, heightened global tensions, resource depletion, and negative health consequences. As a result, prominent environmental organisations and governments have risen to the challenge, adopting programmes such as the Paris Agreement to implement action plans to lower carbon emissions and battle climate change. Despite the various environmental and performance benefits of currently available commercial electric vehicles, electric vehicles only account for a small portion of the automobile market. Consumer interest in electric vehicles (EVs) is waning due to greater pricing, a restricted driving range, and a lack of supporting infrastructure when compared to conventional internal combustion vehicles.

In order to reduce the cost of the EV's battery pack, the greater cost is a vital issue to consider. Machine learning is a low-cost, time-saving way of identifying low-cost, high-performance battery materials and enhancing battery manufacturing efficiency. Artificial intelligence (AI) algorithms and controls can estimate actual driving ranges and optimise energy conservation, allowing for more driving range and lowering customer range anxiety.

However, a number of obstacles must be overcome before such benefits can be realised. Because of the limited range and costs associated with charging EV batteries, it is critical to design algorithms that reduce costs while also preventing users from becoming stranded. In today's world, AI algorithms provide simpler and more comfortable solutions. We end our study by outlining our expectations for the discipline in the future, and the research opportunities that remain open to both the corporate and academic sectors.

Keywords: Electric Vehicles; Alternative Energy; Battery Pack; Station Smart Grid; Intelligent Systems.

3.1 Introduction

We all know that the electric vehicle revolution is accelerating around the world, and there are a variety of models on the market. Within the next 5 years, EVs are predicted to reach price parity with traditional combustion-engine vehicles, thanks in large part to the involvement of artificial intelligence. By hastening improvements in battery technology, AI is assured to give electric vehicles a boost. According to global market projections, between 2020 and 2026, the automotive market is expected to grow at a CAGR of more than 35%. Many electric vehicles are being tested for self-driving to gather data, analyse, and repair EVs. Artificial intelligence is commonly employed in manufacturing and assembly lines. As per a recent study conducted by the European Union (EU), the transportation sector accounts for around 28% of overall CO_2 emissions. Road transport accounts for nearly 70% of total emissions [1].

3.2 Advantages of Electric Vehicles

- **Zero emissions:** No tailpipe pollutants, such as CO_2 or nitrogen dioxide, are produced by this type of vehicle (NO_2). Even though making batteries has a negative effect on carbon emissions, the methods of making batteries are better for the environment.
- **Simplicity:** A lower number of components in an EV engine means lower maintenance expenses. They do not need a cooling circuit, nor do they need a gearshift, clutch, or noise-reduction parts.
- **Reliability:** These cars are less prone to breaking down since they feature fewer and simpler components. Also, EVs do not get worn down by engine explosions, vibrations, or gasoline corrosion, which can happen in regular cars.

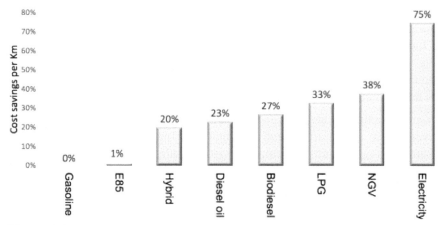

Figure 3.1 Analysing the cost-per-kilometre savings afforded by different fuelled vehicles [5].

- **Cost:** In comparison to ICE, maintenance expenses and the cost of the electricity required are lower. As demonstrated in Figure 3.1, the energy cost per kilometre in EVs is much lower than in regular automobiles.
- **Comfort:** Because there are no tremors or engine noise, EVs are more comfortable to ride [2].
- **Efficiency:** Electric vehicles have a higher efficiency rating than traditional cars. The efficiency of the power plant, on the other hand, will affect the overall well-to-wheel (WTW) efficiency. The entire WTW efficiency of gasoline cars, for example, varies from 11% to 27%, while diesel vehicles' total WTW efficiency varies from 25% to 37% [3]. The WTW efficiency of EVs that run on natural gas is between 13% and 31%, while the WTW efficiency of EVs that run on renewable energy is up to 70%.
- **Accessibility:** This vehicle type allows access to areas where other combustion vehicles are not permitted (e.g., low emissions zones). In large cities, EVs are not subject to the same traffic restrictions as cars, especially when pollution levels are high. Surprisingly, a recent OECD study found that electric vehicles will not improve air quality, at least in terms of PM emissions [4].
- **Range:** On one complete charge, the driving range is limited to 200–350 kilometres. However, this concern is constantly being addressed. The Nissan Leaf, the best example, has a maximum driving range of

364 kilometres [6], whereas the Tesla Model S has a range of more than 500 kilometres [7].

- **Charge time:** Fully charging the battery takes 4–8 hours. "Fast charging" to 80% capacity can take up to 30 minutes. For example, charging time for Tesla superchargers can be 50% in 20 minutes and up to 80% in half an hour [7].

EV manufacturers can now analyse large volumes of sensor data faster than ever before thanks to AI and machine learning, providing them with an unparalleled opportunity to improve existing maintenance operations and potentially add something new: predictive maintenance.

Artificial intelligence (AI) is defined as the ability to comprehend and solve issues that can be created and stored on a computer (or AI). An agent and its surroundings comprise an AI system. An agent is anything that can understand and act on its environment by any means (e.g. a sensor). It is done by studying how human brains work, how people make decisions and work to solve issues, and then applying what they have learned to create software and systems. A large amount of data is used to train the software and achieve the desired results. As an example, when building voice recognition software, a training set containing a variety of data representing various types of speech by various people is employed.

They are the best source of energy in the car in terms of energy density and refuelling time, but strict global regulations on pollutants and CO_2 emissions standards could force humans to abandon fossil fuels in the coming years. Therefore, in this sense, existing CO_2 limits on tailpipes in the most important markets need to be reduced by 34% each year over the next decade. New emission standards need to accelerate technological development to increase efficiency and reduce emissions from internal combustion engines

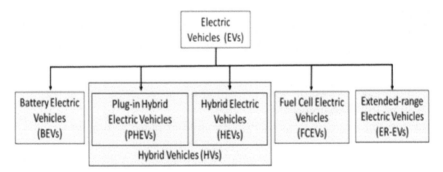

Figure 3.2   Classification of electric vehicles based on engine technologies.

and hybrid vehicles. To this end, AI and machine learning (ML) technologies are increasingly being used in engineering simulation and design software. In the heavy-duty range, increased spark tip pressure, optimised fuel-air mixture and air passage, exhaust energy recovery, lubrication-based friction reduction, innovative piston design, and improved engine control have proven to be an effective approach to breaking the 50% BTE limit. As indicated earlier, 2020 will be a year of growth for electric vehicles. BEVs accounted for 2.8% of all new cars sold in 2019, up from 1.8% in 2018. With BEV sales increasing from 1.1 % in 2019 to 3.6 % in 2020, Europe is leading the way.

Several factors have contributed to the growing sales of BEVs. First, the number of BEV models available in 2020 increased. When you combine this with improved range and lower battery costs, BEVs are becoming a more

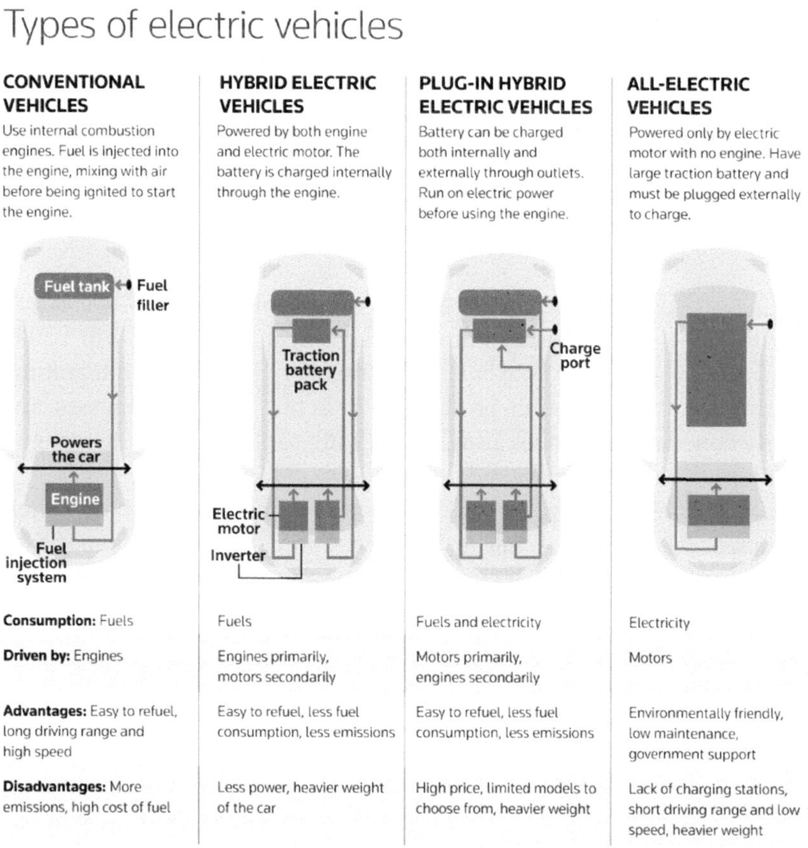

Types of electric vehicles

CONVENTIONAL VEHICLES	HYBRID ELECTRIC VEHICLES	PLUG-IN HYBRID ELECTRIC VEHICLES	ALL-ELECTRIC VEHICLES
Use internal combustion engines. Fuel is injected into the engine, mixing with air before being ignited to start the engine.	Powered by both engine and electric motor. The battery is charged internally through the engine.	Battery can be charged both internally and externally through outlets. Run on electric power before using the engine.	Powered only by electric motor with no engine. Have large traction battery and must be plugged externally to charge.

	CONVENTIONAL	HYBRID	PLUG-IN HYBRID	ALL-ELECTRIC
Consumption:	Fuels	Fuels	Fuels and electricity	Electricity
Driven by:	Engines	Engines primarily, motors secondarily	Motors primarily, engines secondarily	Motors
Advantages:	Easy to refuel, long driving range and high speed	Easy to refuel, less fuel consumption, less emissions	Easy to refuel, less fuel consumption, less emissions	Environmentally friendly, low maintenance, government support
Disadvantages:	More emissions, high cost of fuel	Less power, heavier weight of the car	High price, limited models to choose from, heavier weight	Lack of charging stations, short driving range and low speed, heavier weight

Figure 3.3 Types of electric vehicles (source: US Department of Energy).

EV sales by country

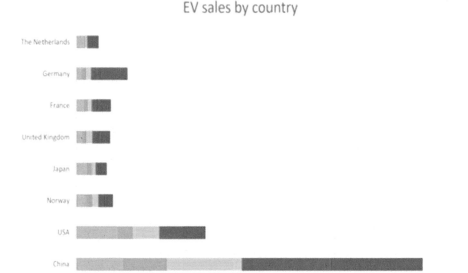

Figure 3.4   The global evolution of electric vehicle scale.

appealing alternative for a larger portion of the population. Artificial intelligence will help the EV business advance significantly since it will make riders' lives easier and more convenient. Automakers are rapidly integrating AI platforms into their existing production systems. These businesses are increasingly concentrating on creating self-driving cars to improve passenger transportation.

Automobiles are evolving, particularly in the EV market, which will alter the experience. AI will wreak havoc by streamlining production capacities and boosting company growth. It can be used for a variety of tasks, including fixing electric vehicles, determining the source of a problem; locating nearby workshops and charging stations; and ensuring the safety of electric vehicles by disconnecting e-bikes or e-cars from mobile devices. EVs, along with shared portability, open transportation and other clever city highlights, will continue to play an important role. As a result, much more effort is needed to make the charging handle simpler and upgrade the battery. The main drawback of electric vehicles is the need for autonomy. Researchers are working on superior battery innovation to boost driving manoeuvres. while decreasing charging time, weight and fetch the long term of electric vehicles will be decided by these contemplations.

3.3 Opportunities for Energy Harvesting on Electric Vehicles

The amount of energy collected in and on electric vehicles for delivery, water and discussion is becoming a multibillion-dollar industry. Include solar-powered ground offices such as aeroplane holders, which charge electric planes and other structures such as the one that will soon be built at Boston Logan Airport to charge Ground Bolster Gear GSE, which includes buses, all of which are going electric. The sun-oriented boards on a Brisbane transport warehouse charge the plan line 100% electric buses nowadays, and a few other buses have sun-oriented boards to offer additional control. Vitality gathering clearly has a financial advantage nowadays, and there will be much more in the future. Creators that disregard these conceivable outcomes will be left behind.

Energy harvesting has gained popularity in recent years as a means of meeting the world's ever-increasing energy demand while also addressing the growing threats of climate change and pollution. Its widespread adoption, similar to that of EVs, provides numerous benefits while also posing numerous challenges. Nonetheless, encouraging the use of solar energy and electric vehicles is likely one of the best ways to address some of the world's challenges.

3.4 The Electric Vehicle Industry and Artificial Intelligence

Battery execution may make or break the EV experience, from the drive run to the charging time to the car's lifetime. Ideas like recharging an electric vehicle in the time it takes to stop at a gas station are now more likely to come true, and this may also help improve other parts of battery innovation.

3.5 How AI Is Accelerating the Power of Electric Vehicle Batteries

The emergence of AI has thrown the electric vehicle ecosystem into disarray. In this article, we will look into how artificial intelligence is used in EV pipelines. Battery performance is an important consideration in every element of a vehicle's life. The battery accounts for 25% of the total cost of an electric car, and nothing is more crucial than boosting battery life when it comes to EV advancements. Artificial intelligence has helped dreams come true, such as charging an electric vehicle in the time it takes to stop at a petrol station, and it may also assist in improving other aspects of battery technology. The global battery market is rapidly expanding in size. EV vehicles such as the Tesla S 100D have a range of 355 miles, the Hyundai Kona has a range of 198 miles and the MG ZS EV has a range of 214 miles, although none of them can be fully charged in under an hour. For example, EVs will take 75 minutes to achieve full charge at a Tesla supercharging station, and Indian EV drivers will need more than 3 hours to reach full charge.

The underlying potential and opportunity of battery life cycle management can be unlocked by machine learning. Data are the key to extending battery life. Machine learning combines modern electronics with IoT, data science and digital twins to predict battery life, identify potential degradation/breakdown and its causes, and prevent delays/errors. ML adds a layer of intelligence after a lot of data has been collected and tracked about the battery's life, performance, state of charge, stress from rapid acceleration and deceleration, temperature, number of charge cycles and other things.

Using in-vehicle operating software is a completely different and exciting application that continuously checks battery performance and condition and feeds back data to drive product innovation. In addition, the development of smarter batteries with built-in sensing and self-healing capabilities will allow battery management systems to recognise their "state" and activate battery cells or modules as needed. AI helps innovators gain insights not possible with traditional statistical analysis. This lets you change more than one part at once and lets you check the proof faster.

3.6 Internal Combustion Engine (ICE) Sales Ban Proposals

Country	Current government proposals to ban ICE only vehicle sales
China	Actively considering and studying a ban
France	2040
Germany	2030
India	2030
Ireland	2030
Israel	2030
Netherlands	2030
Norway	2025
Scotland	2032
UK	2040

Figure 3.5 ICE sales ban proposals for different countries by their government (source: Thomson Reuters GFMS team).

3.7 Electric Vehicle Charging

According to the Electric Power Research Institute, electric vehicle charging is classified into three stages. Level 1 charging, which operates at 120 V/15 A [8], is the slowest. In this instance, the charging equipment is mounted on the electric vehicle, and power is transmitted to it via a plug and cord set. Level 2 charging, on the other hand, operates at 240 VAC and has been used in both private and public settings. It charges at a faster rate than level 1, although it necessitates the purchase of specialised charging equipment. Level 3, often known as "rapid charging," uses 480 VAC and is commonly used in commercial and public venues to provide "grab and go" service for ICE vehicles, comparable to petrol stations. Level 3 charging is the fastest, with automobiles charging in under 30 minutes. The three charging strategies are depicted in Figure 3.6.

Residential and non-residential EV charging can also be distinguished. For household use, level 1 and level 2 chargers are usually used. Residential

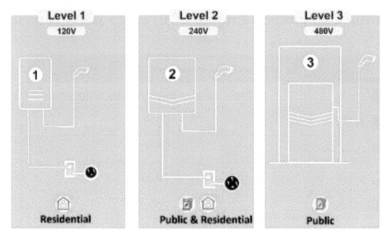

Figure 3.6   Levels of EV charging.

billing patterns are more predictable, making scheduling more straight-forward. Users typically charge their automobiles overnight or schedule charging sessions around their work schedules. The number of vehicles charging at home is also predictable since people who own electric vehicles in a certain location are more likely to use charging stations in that area. The number of cars that use non-residential charging stations, on the other hand, is hard to predict because it depends on many things.

3.8 ML and Predictive Analytics

ML allows computers to learn from their mistakes without having to program them explicitly. The dataset that the algorithms used to train themselves is called "experience" in this context. With enough time, the models can uncover the dataset's underlying trends and patterns. These models can make accurate future predictions and so provide predictive analytics after successful learning. Supervised and unsupervised learning are two types of machine learning algorithms.

It can be further categorised according to the type of variable expected (also known as the response variable). If the response variables are continuous, the problem to be solved is called a regression problem. This problem is called a classification problem. The response variable is categorical; in view of EV charging, Figure 3.7 shows the difference between regression and classification. The graph on the left shows a forecast of energy consumption based on charge time. This is a regression problem because the

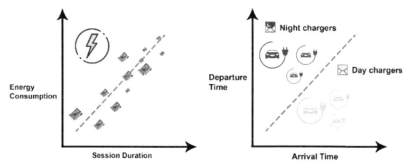

Figure 3.7 Regression (left) and classification (right) difficulties are depicted.

energy response variables are continuous. On the other hand, the image on the right shows the difference between an EV driver who prefers to charge the car at night and an EV driver who prefers to charge the car during the day.

3.9 Supervised Learning

In supervised learning, ML models are trained using a labelled training dataset. As a result, the dataset includes both the input variables and the target variable, also known as the response variable. The model learns the mapping between the input and response variables by iteratively optimising a given objective function. For example, in the context of EV charging, a dataset containing the vehicle's arrival time, city name and departure time, for example, could be a simple example. If the goal is to predict the time of departure, the ML model will learn the relationship between the time of arrival, the name of the city (both of which are input variables) and the time of departure (which is the response variable).

3.10 Unsupervised Learning and Statistical Models

The training dataset for unsupervised learning contains just input variables and no labelled output variables. Data labelling is time-consuming and costly in many real-world applications. The ML model's purpose is to detect patterns or structures in the data. Cluster analysis is a popular type of unsupervised learning in which the purpose of the machine learning model is to locate clusters of items that share some common characteristics. Finding clusters of EV behavioural patterns can be done via unsupervised learning. Figure 3.8 clustering in the context of electric vehicle charging. In this straightforward example, only two input features are used to group the objects. We can

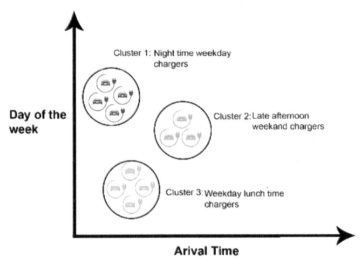

Figure 3.8   Simple illustration of clustering.

identify three distinct charging behaviour groups based on the arrival time and day of the week.

3.11 Electric Vehicle Battery Makers

Electric vehicles are powered by lithium-ion (Li-ion) cells. The battery is the most expensive component in an EV, accounting for 40–50% of the total cost. These batteries have revolutionised how goods are designed in today's society.

Lithium-ion batteries are being heralded as the future of car energy, and properly so, given their inherent benefits. For example, lithium-ion batteries are substantially lighter than other rechargeable batteries and keep their charge exceedingly well while delivering steady energy production. The rising use of electric vehicles in India will increase the demand for lithium-ion battery production. China is the largest supplier of lithium-ion cells in the world, so most companies that make electric cars import cells and batteries from there.

On the other hand, companies are speeding up plans to manufacture lithium-ion cells in the country, expecting to take advantage of Rs 180 billion in government subsidies. Electric vehicle battery recycling will increase in popularity due to the scarcity of rare minerals like cobalt and lithium, as well as environmental concerns. According to certain projections, the worldwide electric car battery recycling industry would reach $2.27 billion by 2025, with a 41.8% compound annual growth rate. If all you have is a hammer, every problem appears to be a nail. When considering the rush to promote artificial intelligence for nearly every system and flaunt machine learning as a silver bullet to address whatever value the new product or service is said to deliver, this overly-user cliché seems extremely apt. For the next 5–10 years, internal combustion vehicles will continue to dominate the new vehicle market. According to Bloomberg New Energy Finance (BNEF), EVs will account for 44% of all new vehicle sales in Europe by 2030, 41% in China, 34% in the United States and 17% in Japan.

Due to a lack of charging infrastructure and cheap EV models, India would lag behind these regions, accounting for only 7% of total sales. In order to accelerate the adoption of electric vehicles in India, the government plans to deploy up to 70,000 EV chargers over the next few years. The work has already begun. Demand for resources such as lithium, cobalt, copper, and nickel is already increasing as EV demand grows. There could be a supply shortage because these metals are limited and are the most commonly used metals in the production of electric car batteries. Hindustan Copper is placing a large bet on electric vehicles to boost demand growth. The production of electric vehicles will use four times as much copper as typical internal combustion engines. This is good news for Hindustan Copper, NALCO, and Hindalco, among others. Copper and aluminium will be the main materials used in EVs, so the switch from internal combustion engines to EVs will help these industries a lot.

The most important properties of the various technologies introduced are compared in Figure 3.9. The working temperature of the various technologies is a crucial factor to consider when comparing them, as it can limit their acceptance. In this regard, lead-acid and lithium batteries are the most resistant to low temperatures, as they can withstand load temperatures of up to 200 °C, albeit low temperatures severely reduce the capacity of Li-ion batteries, causing self-discharge [9]. The ideal operating temperature for this type of battery is 400°C.

The advancement of lithium-ion battery technology has been critical to this expansion. As Jon Berntsen, a senior energy analyst at Thomson Reuters, explains, technological advancements in this field are leading to a plummet in energy storage prices.

"The price of a lithium-ion battery pack is constantly reducing by 15% every year due to economies of scale, and the energy density is growing," Berntsen continues. "As a result, the range is extended at the same price."

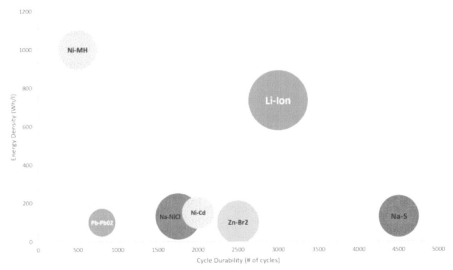

Figure 3.9   A comparison of battery systems based on their cycle durability and energy 'density.

Consumers will embrace EVs as the range rises, and adoption will follow a conventional technology adoption curve, from early adopters to laggards, which is similar to others in the technology sector in the market.

3.12 Is Lithium the New Gasoline?

Switching to a zero-emission vehicle is not a zero-sum game. As the sales of electric vehicles increase, the demand for alternatives to clean energy increases, such as lithium, used for batteries to store electricity, silver and copper used for solar power generation and charging infrastructure, and rare earth. The demand for various other materials is increasing with an electric motor etc.

According to the GFMS Metals team of Johann Wiebe and Thomson Reuters, demand for lithium, cobalt, nickel, rare earths and graphite will grow the most in a world with 100% EV penetration.

Based on the rising use of lithium-ion battery technology in the battery electric car sector, some have dubbed lithium "the new gasoline." Lithium accounts for 12% of battery costs, and the electric car sector now accounts for roughly 14% of lithium demand. By 2025, that percentage is expected to rise to 40%. But lithium is not the only element involved. Cobalt and rare

Metals demand in a 100% EV world

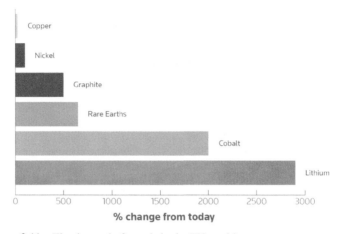

Figure 3.11   The demand of metals in the EV world.

earths will be in high demand, and the real environmental and geopolitical effects will need to be carefully watched.

3.13 Enabling High-Energy Dense Batteries

High-energy density lithium-ion batteries allow EVs to enhance cargo capacity and range while lowering costs—an important factor in the EV market's continued growth. On the other hand, cooling these batteries has made it very hard to make them even more powerful.

The downside of high energy-density lithium-ion batteries is that they produce more heat over the same surface area as their predecessors. "Cooling more powerful battery cells without having the extra surface is always a challenge," Vanton says. "Cooling must then be much more efficient and uniform." A traditional s-shaped thermal design will not be sufficient in circumstances where the pressure drop must be essentially constant.

The AI-driven method is breaking down these cooling barriers. A Diabatix designed liquid cooled heat sink and a typical liquid cooled heat sink for an EV battery are shown in the following comparison. In terms of outcomes, our design outperforms the competition.

In Diabatix's design,

1. Thermal resistance was decreased by 25%.

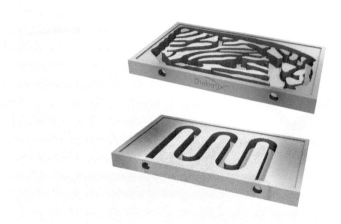

Figure 3.12 Diabatix designed liquid-cooled heat sink and a typical liquid-cooled heat sink for an EV battery (source: https://www.diabatix.com/blog/ai-driving-better-ev-cooling).

2. Temperature uniformity was improved by 15%.
3. Pressure drop was reduced by 50%.

Designs that generate improved performance to cool higher-powered batteries without having to increase surface area are produced as a result of the AI-driven technique. This is particularly important in terms of material prices.

3.14 Future Scope of AI in the EV Industry

Despite the fact that the development and evolution of electric vehicles have accelerated in recent years, we comment in this section on elements that are still unresolved or that may be worth investigating in order to provide new and improved solutions.

AI is playing a crucial role in the EV industry, with applications like autonomous driving, user behaviour monitoring, and smart navigation systems. It can be utilised for a variety of safety applications, including equipment predictive maintenance, driver behaviour monitoring, and vehicle security. A variety of organisations are using artificial intelligence in their vehicles; some are using it to replace present transportation systems with self-driving services, while others are using it to improve the battery power of their electric vehicles.

Other features, however, are equally important, but they must be improved if electric vehicles are to be used. The first of these factors is the global deployment of EV charging stations. The majority of countries currently have a scarcity of charging stations, which deters purchasers. We believe that more effort is required to strengthen the infrastructure for charging. Furthermore, the time it takes to fully charge the batteries in these vehicles must be significantly lowered, making electric vehicles more appealing to consumers. We think that putting AI and vehicle communications together can help speed up the spread of new, more sustainable and environmentally friendly ways to get around.

There are numerous AI-based total answers for numerous EV areas, which include electricity green routing, higher and smarter charging, and battery temperature control. In their study, Masikos *et al.* [10] describe a unique system gaining knowledge of-primarily based totally technique for electricity green routing. Their method may also expect electricity utilisation for exceptional street segments that make up modern-day or destiny car routes. Alesiani and Maslekar [11] speak about the problems of growing routes for a fleet of electrical vehicles. Their answer considers now no longer the most effective car's battery capacity, however, additionally using charging stations alongside the route, all at the same time as the usage of a developing genetic set of rules with gaining knowledge of technique.

Sugii *et al.* [12] offer a genetic algorithm-based scheduling technique for charging numerous EVs as part of smarter charging. This method can be used to figure out the power curve and electric power load levering in particular. It can also lower the charging equipment's starting cost and capacity. Because user behaviours are one of the most crucial challenges in EV charging, Panahi *et al.* [13] propose using ANNs to estimate the daily load profiles of individual EVs and fleets. They use previous data in particular to forecast electricity consumption and better manage charges. Park et al. [14] suggest using artificial neural networks (ANNs) to improve the thermal management system and minimise total energy consumption in the context of battery thermal management. The proposal enables the battery temperature to be kept within a reasonable range. The relationship between battery thermal behaviour and design characteristics is investigated by Karimi *et al.* [15]. Their numerical research shows that at varied discharge rates, a cooling technique based on distributed forced convection may offer homogeneous temperature and voltage distributions throughout the battery pack.

Almost every major automaker is considering making the switch to electric vehicles, and Europe, China and India are on the edge of making fossil-fuel vehicles illegal to sell. We are also seeing the end of the linear growth curve for hybrid vehicle sales and the beginning of the exponential growth curve for electric vehicle sales. Transportation-related carbon emissions reduce as a result, while electricity demand and grid stress rise dramatically. By 2040, almost half of all vehicles on the road will still be powered by fossil fuels, but all new vehicles sold will be electric. As a result, carbon dioxide emissions from passenger cars will fall to 1.7 billion metric tonnes, but the total energy required to power the world's increasingly electric fleet of vehicles will rise to almost 1,350 terawatt hours.

The rapid evolution of the automobile industry will have an impact on practically every other industry on the planet. Major changes are expected in energy production, metals and mining, global trade and transportation, technology, environmental legislation, tax credits and incentives. The challenge for businesses functioning in this environment is not so much that change is taking place, but that it is taking place at an unpredictable and frequently unequal rate.

Improved wireless charging methods and higher-capacity batteries will make it easy to use the fastest and most powerful charging modes. Another aspect that could help electric car adoption is the creation of a one-of-a-kind connector that can be used globally. In the future, electric vehicles will play a significant role in smart cities, and having a range of charging systems that can respond to user needs will be critical. While many people feel that personal, self-driving automobiles are the way of the future, AI and machine learning are currently being employed in a variety of ways in vehicle design and operation. Smart electric vehicles from corporation enable the creation of more efficient and sustainable cities by freeing up parking, reducing the number of cars on the road, and lowering environmental effects.

3.15 Conclusion

ML in R&D is not the only way AI may aid in the advancement of electric vehicles. Implementing it within a vehicle's operational software, where it would continuously monitor battery performance and health and feed data back to improve product innovation, would be a completely unique and exciting application. Furthermore, by creating smarter batteries with embedded sensing and self-healing capabilities, the battery-management system will be aware of their state of health and will be able to rejuvenate battery cells or modules as needed.

These results cannot be obtained through traditional statistical analysis. It allows them to tweak more than one component at a time and analyse evidence more quickly. This evidence enables speedier development cycles and the resolution of problems that would otherwise be unsolvable. This capacity is critical in overcoming one of the most significant consumer barriers to EV adoption: "range anxiety." If machine-learning technology can shorten the time it takes to charge batteries, it could change the electric vehicle business as a whole.

A comprehensive approach to data science in battery development could be the key to resolving such complex models. Artificial intelligence, also known as machine learning, is capable of analysing data and constructing mathematical models at a far faster rate than the human brain. Without having to be explicitly coded, systems can learn and improve on their own. The current and potential impact of AI on a variety of businesses is enormous.

References

[1] European Commission (2011), Transport in Figures—Statistical Pocketbook. Available online: https://ec.europa.eu/transport/f acts- funding's /statistics/pocketbook-2011_en.

[2] Chan, C. C. (2007), the state of the art of electric, hybrid, and fuel cell vehicles. Proc. IEEE, 95, 704–718.

[3] Albatayneh, A.; Assaf, M. N.; Alterman, D.; Jaradat, M. (2020), Comparison of the Overall Energy Efficiency for Internal Combustion Engine Vehicles and Electric Vehicles. Environ. Clim. Technol. 24, 669– 680.

[4] OECD iLibrary (2020), Non-Exhaust Particulate Emissions from Road Transport: An Ignored Environmental Policy Challenge; Technical Report; OECD Publishing: Paris, France. Available online: https://do i.org/10.1787/4a4dc6ca-en.

[5] Blázquez Lidoy, J.; Martín Moreno, J. M. (2010), Eficiencia energética en la automoción, el vehículo eléctrico, un reto del presente. Econ. Ind.377, 76–85.

[6] Nissan. Nissan Leaf. Available online: https://www.nissan.co.uk/vehic les/new-vehicles/leaf/range-charging.html.

[7] Tesla (2019). Tesla Official Website. https://www.tesla.com/en_EU/su percharger

[8] K. Morrow, D. Karner, and J. Francfort, (2008), "Plug-in hybrid electric vehicle charging infrastructure review," U.S. Dept. Energy, US Dept. E nergyVehicle Technol. Program, Contract No. 58517, vol. 34.

[9] Du Pasquier, A.; Plitz, I.; Menocal, S.; Amatucci, G. (2003). A omparative study of Li-ion battery, supercapacitor and non-aqueous asymmetric hybrid devices for automotive applications. J. Power Sources, 115, 171–178.

[10] Masikos, M.; Demestichas, K.; Adamopoulou, E.; Theologou, M. (2013) Machine-learning methodology for energy efficient routing. IET Intell. Transp. Syst. 8, 255–265.

[11] Alesiani, F.; Maslekar, N. (2014), Optimization of Charging Stops for Fleet of Electric Vehicles: A Genetic Approach. IEEE Intell. Transp. Syst. Mag. 6, 10–21.

[12] Sugii, Y.; Tsujino, K.; Nagano, T. (1999), A genetic-algorithm based scheduling method of charging electric vehicles. In Proceedings of the IEEE International Conference on Systems, Man, and Cybernetics (Cat No. 99CH37028), Tokyo, Japan, 12–15; Volume 4, pp. 435–440.

[13] Panahi, D.; Deilami, S.; Masoum, M. A. S.; Islam, S. M. (2015), Forecasting plug-in electric vehicles load profile using artificial neural networks. In Proceedings of the 2015 Australasian Universities Power Engineering Conference (AUPEC), Wollongong, Australia, pp. 1–6.

[14] Park, J.; Kim, Y. (2020), Supervised-Learning-Based Optimal Thermal Management in an Electric Vehicle. IEEE Access, 8, 1290–1302.

[15] Karimi, G.; Li, X. (2013), Thermal management of lithium-ion batteries for electric vehicles. Int. J. Energy Res., 37, 13–24.

4

Advances in Maximum Power Point Tracking of Solar Photovoltaic Systems Under Partially Shaded Conditions with Swarm Intelligence Techniques

Dileep Krishna Mathi[1] and Ramulu Chinthamalla[2]

[1]EEE Department, Vidya Jyothi Institute of Technology, India
[2]EE Department, National Institute of Technology, Warangal, India
E-mail: dileepkrishnaeee@gmail.com; rnitchinthamalla@nitw.ac.in

Abstract

Abundant solar energy is an alternative to fossil fuels to produce green and clean electricity for high energy demands. The popularity of solar energy is limited by its high initial costs and the lower conversion efficiency of photovoltaic (PV) modules. The PV system's maximum energy yield at any atmospheric conditions is essential to get back the high initial costs. For this purpose, several maximum power point tracking (MPPT) techniques are widely used. The output of the PV panels is adversely affected due to atmospheric conditions like partially shaded conditions (PSC), and even these can physically damage the PV cell. Hence, to protect PV panels from physical damage, the traditional method is to connect a bypass diode across the group of cells. But this leads to a multi-peak power versus voltage (P–V) curve, which makes it difficult to track the maximum power point (MPP) with conventional MPPT techniques. Hence, advanced optimisation techniques are used to find the optimum value on the multi-peak P–V curve under PSC. However, these techniques fail to track GMPP due to improper algorithm design, which leads to high energy losses. In this chapter, MPPT techniques for PV systems with PSC are looked at, and new control strategies for efficient MPPT tracking are introduced.

Keywords: Maximum Power Point Tracking (MPPT); Solar Photovoltaic Systems; Partially Shaded Conditions (PSC); Global Peak (GP).

4.1 Introduction

In recent years, the electrical energy demand has sharply increased due to urbanisation and in-deserialisation in most countries. There are significant shifts in energy demand, which are predicted by the international energy agency (IEA) and demand is growing faster in developing countries. Nowadays, increasing interest in green energy and the depletion of fossil fuels has increased the use of renewable sources of energy to decrease global warming. Hence, to meet the increase in the load demand of developing countries like India, solar energy has become the major renewable energy source [1, 2].

The main reasons for the popularity of solar energy among other renewable energy sources are that there is plenty of free and abundant solar energy. Converting it to electrical energy with photovoltaic (PV) panels is simple and leads to the absence of moving or rotating parts [3, 4], as shown in Figure 4.1.

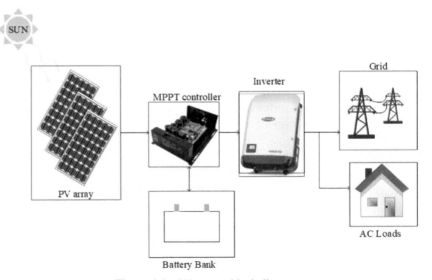

Figure 4.1   PV system block diagram.

Where the expansion of the grid is not feasible, it is possible to generate power from solar energy in remote areas.

PV panels/modules convert solar energy to electrical energy in PV system [5]. These systems are classified as on-grid and off-grid PV systems based on the connected load.

4.2 Model Description of PV Source

The PV panel/module is the semiconductor device used to convert photon energy into electrical energy. A PV cell is a basic building block of the PV panel/module. Several cells are connected in series and in parallel in a PV panel. Its mathematical model is given by eqn. (4.1), and each cell in the module can be represented with an equivalent circuit consisting of a single diode, which is known as the single diode model, as shown in Figure 4.2 [6]:

$$I_{pv} = I_{ph} - I_0 \times \left(e^{\left(\frac{V_{pv} + I_{pv} \times R_s}{N_s * a * V_T} \right)} - 1 \right) - \frac{V_{pv} + I_{pv} \times R_s}{R_{sh}}, \qquad (4.1)$$

where I_{pv} is the PV panel/module current, I_{ph} indicates the photo-current, I_0 indicates the saturation-current of the diode, V_{pv} represents the PV panel terminal voltage, R_s represents the series resistance, N_s indicates the number of cells in series, a indicates the ideality of the diode, V_T represents the diode thermal voltage and R_{sh} indicates the shunt/parallel resistance.

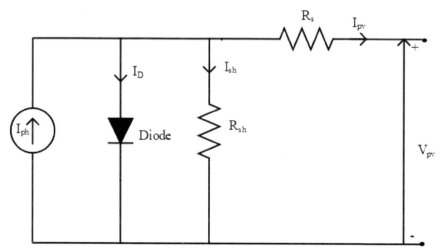

Figure 4.2 Electrical equivalent circuit diagram of PV panel by using single diode model.

Based on the system's voltage requirement, several modules are connected in a series arrangement known as a string. In an array, several strings are connected in parallel to get the required output power from the PV system, making the PV array as shown in Figure 4.3.

For example, a 300 W PV system can be formed with 50 W solar panels manufactured by the Solar Power Mart SPM-50M model. Its electrical characteristics are given in Table 4.1, with six panels connected in either six series one parallel (6S-1P) or two 3-panel strings connected in parallel (3S-2P short string) as shown in Figure 4.4. To get a higher voltage level at the PV terminals, the number of modules in the series string is increased. The

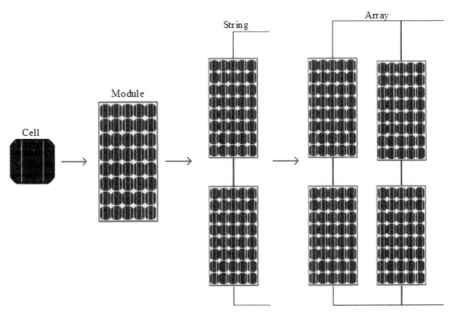

Figure 4.3   PV cell to array configuration.

Table 4.1   Electrical characteristics of solar power mart SPM-050M model 50 WP PV panel at standard test conditions AM 1.5.

S. no	Parameter	Value
1	PV maximum power P_{mp}	50 W
2	Voltage at power maximum V_{mp}	18.68 V
3	Current at power maximum I_{mp}	2.68 A
4	Open-circuit voltage V_{oc}	22.32 V
5	Short-circuit current I_{sc}	2.86 A
6	Temperature coefficient of P_{mp} ()	$-0.45 \pm 0.05\%$ °C

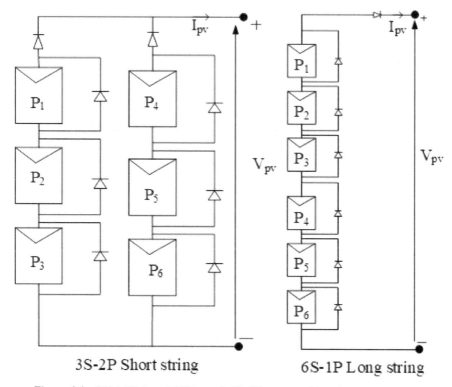

3S-2P Short string   6S-1P Long string

Figure 4.4 SPM-50M model PV panels P1–P6 connected in string arrangement.

electrical equivalent circuit parameters of the given solar panels are extracted using an analytical method proposed in [7] and values obtained are series resistance R_s = 0.24 Ω, shunt/parallel resistance R_{sh} = 613.6 Ω and diode saturation current I_0 = 8.46 × 10^{-10} A.

The popularity of PV systems is limited by the intermittent nature of PV sources during low insolation periods, that is, during nights or under partially shaded conditions (PSC) and nonlinear current versus voltage (I–V) characteristics, which leads to the single maximum power point (MPP) where optimum power can be extracted. Other drawbacks of PV systems are the high initial cost and low conversion efficiency of the solar panels [8]. It is imperative to extract the optimum PV power from the system under varying weather conditions to get the profit or the high initial cost of the PV system. These atmospheric conditions are classified as uniform or nonuniform based on the irradiance of each panel in the PV array, as shown in Figure 4.5.

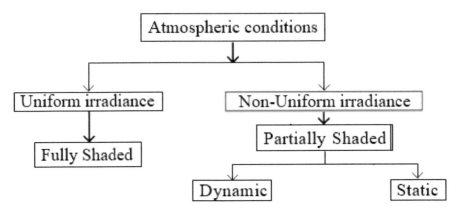

Figure 4.5   Classification of atmospheric conditions across PV system based on irradiance.

4.3 Partial Shading and Its Effects

The main differences in irradiance at the PV modules of string are due to the shade of buildings, trees, clouds, ice formation, bird droppings and dust formation, as shown in Figure 4.6, which is treated as a shading condition. The photocurrent generated in the shaded cell is reduced by 20% compared to un-shaded cells. Hence, shaded modules affect the string's total power under PSC and power loss due to shading is as high as 40% [6]. In order to operate at the higher current level of an unshaded module in the string, the shaded module's operating point is shifted to the negative voltage region. Hence, these cells dissipate power and are physically damaged permanently due to hot spots.

For mitigating the power loss due to partial shading, the following methods are shown in Figure 4.7 are found in the literature. PV array reconfiguration and PV system architectures are costly and complex in nature. The easiest solution is to bypass the string current of the shaded module by using the bypass diode across the group of cells in the module, as shown in Figure 4.8.

Since the power produced in the shaded module is bypassed, this loss is considered mismatch loss, as shown in Figure 4.9, and the power versus voltage (P–V) of the array becomes a multi-peak curve under PSC, which has a single global peak (GP) and multiple local peaks, whereas the P–V curve of the array is a single-peak curve under uniform irradiance conditions (UIC). Most of the conventional maximum power point tracking (MPPT) techniques

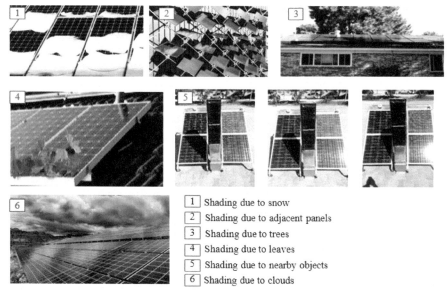

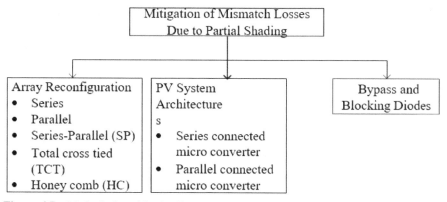

1	Shading due to snow
2	Shading due to adjacent panels
3	Shading due to trees
4	Shading due to leaves
5	Shading due to nearby objects
6	Shading due to clouds

Figure 4.6 Partial shading due to various atmospheric conditions.

```
                    ┌──────────────────────────────┐
                    │  Mitigation of Mismatch Losses │
                    │     Due to Partial Shading     │
                    └──────────────────────────────┘
```

Array Reconfiguration	PV System Architecture s	Bypass and Blocking Diodes
• Series • Parallel • Series-Parallel (SP) • Total cross tied (TCT) • Honey comb (HC)	• Series connected micro converter • Parallel connected micro converter	

Figure 4.7 Methods found in the literature for the mitigation of mismatch losses due to partial shading.

are best suited for single-peak P–V curves, whereas it is difficult to track the GP of a multi-peak P–V curve under PSC with the same techniques. If the MPPT technique fails to track the GP, the power loss is considered a power loss due to MPPT failure.

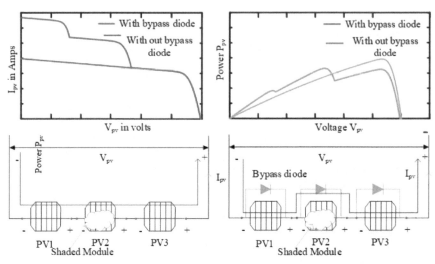

Figure 4.8   Current flow in a shaded module without and with bypass diodes.

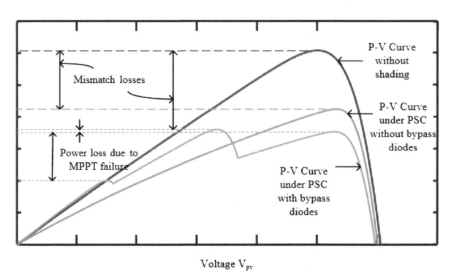

Figure 4.9   Power losses due to shading.

4.4 Model Description of MPPT Controller

Fuller et al. proposed a practical PV system in 1954, as shown in Figure 4.10 4.10, in which DC load can be connected directly to the PV panel. From

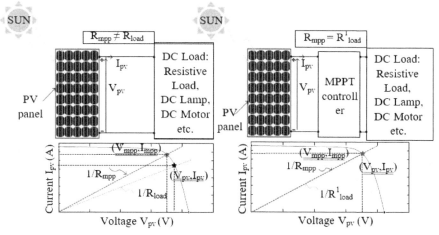

Figure 4.10 PV panel connected to load.

Figure 4.10, it will not ensure the operation at MPP, which varies dynami-cally with the change in insolation and temperature. As a result, researchers propose the maximum power point tracker, a power electronic converter for medium and high power-rated PV systems [9]. Given the low conversion efficiency of PV panels, it is important to increase the efficiency of the PV system. The use of maximum power point trackers increases the PV system efficiency by 30% as compared to the system without MPPT. In recent years, new grid codes also impose maximum power point tracking for all grid-connected systems to get maximum energy yield and active power limit control to maintain stability under peak power production periods, that is, during sunny days [10]. So, both stand-alone and grid-connected PV systems should have an MPPT controller to get the most power out of the solar panels in all kinds of weather.

The MPPT converter is a power electronic converter that adjusts the effec-tive load resistance at PV array terminals by varying the switches' control signals. In general, a DC–DC converter acts as an MPPT converter in a two-stage power conversion system, whereas the inverter itself acts as an MPPT converter in a single-stage power conversion system. In this thesis, the main objective is to verify the performance of various MPPT techniques; only the DC–DC conversion stage of a two-stage power conversion system is used, as shown in Figure 4.11. Based on the load voltage requirement, a proper DC–DC converter is used among the basic DC–DC converter types, that is, boost, buck and buck–boost. In the case of a battery charging application, the PV

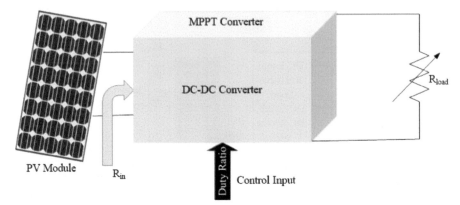

Figure 4.11   DC–DC converter as an MPPT converter.

input voltage should be scale down to the battery voltage level. Hence, a buck converter is used as an MPPT converter for battery charging applications. For the system-independent online PV MPPT, voltage and current at the terminals of the PV are sensed and given as input to the MPPT algorithm, as shown in Figure 4.12. Based on the logic used in the algorithm, it generates the

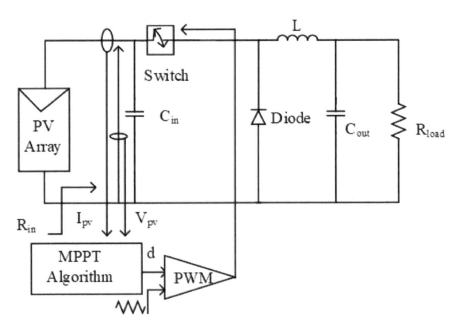

Figure 4.12   DC–DC buck converter as an MPPT converter.

control signal. Input resistance seen at the terminals of the PV array is given by eqn. (4.2)

$$R_{in} = \frac{R_{load}}{d^2},$$ (4.2)

where R_{in} is the PV input resistance, R_{load} represents the load resistance and d indicates the duty ratio. The operating characteristics of the buck converter are shown in Figure 4.13. When duty ratio (d) ranges from zero to one, R_{in} varies from ∞ to R_{load}. As per the maximum power transfer principle, maximum power is transferred to the load when the load resistance is equal to PV resistance at maximum power ($R_{mp} = R_{load}$), where $R_{MP} = \frac{V_{mp}}{I_{mp}}$ is the resistance at the MPP. If the load resistance is greater than R_{mp}, MPP will not be covered in the operating region of the buck converter. Hence, R_{load} should be smaller than R_{mp} to track the MPP with a buck converter.

To boost the low input PV voltage to a higher level, a boost converter is used as an MPPT converter, as shown in Figure 4.14. In the case of a grid-connected solar PV system, the PV input voltage should be scaled up to the required DC-link voltage. Hence, a boost converter is used as an MPPT converter for the grid-connected PV system. Input resistance seen at the PV array terminals is given by eqn. (4.3)

$$R_{in} = R_{load} \times (1-d)^2.$$ (4.3)

The operating characteristics of the boost converter are shown in Figure 4.15. When duty ratio (d) ranges from 0 to 1, R_{in} varies from R_{load} to

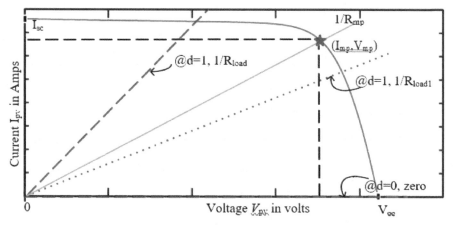

Figure 4.13 Operating region of a buck converter.

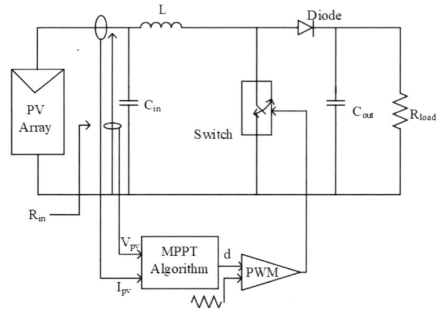

Figure 4.14   Boost converter as an MPPT converter.

zero. As per maximum power transfer principle, when $R_{mp} = R_{load}$ maximum power is transferred to the load. If the load resistance is smaller than R_{mp}, MPP will not be covered in the operating region of the boost converter. Hence, R_{load} should be greater than R_{mp} to track the MPP with the boost converter.

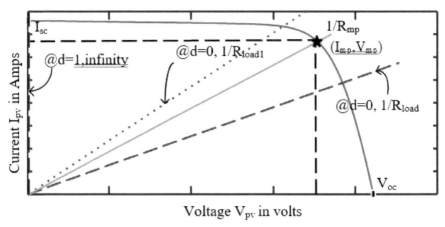

Figure 4.15   Operating region of a boost converter.

To get the MPP tracking without independent of load value, a buck–boost converter is used as an MPPT converter as shown in Figure 4.16. Input resistance seen at the PV array terminals is given by eqn. (4.4)

$$R_{in} = R_{load} \times (\frac{1-d}{d})^2. \tag{4.4}$$

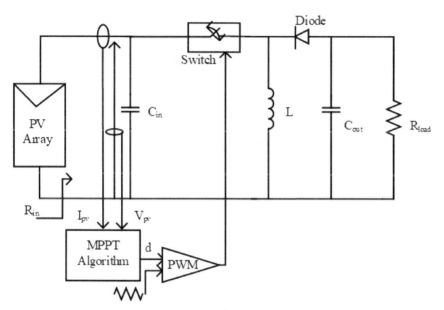

Figure 4.16 Buck–boost converter as an MPPT converter.

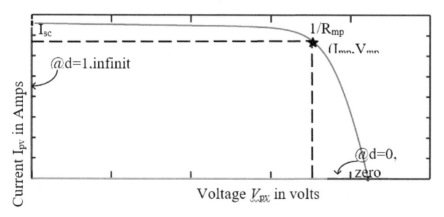

Figure 4.17 Operating region of a buck–boost converter.

When duty ratio (d) ranges from 0 to 1, R_{in} varies from infinity to zero. As per the maximum power transfer principle, when $R_{mp} = R_{load}$ maximum power is transferred to the load. There is no restriction in choosing the load resistance value.

4.5 Overview of MPPT Techniques for Solar Photovoltaic Systems

The MPPT controller uses an algorithm to generate the control pulses based on the inputs of the controller. First, Fox et al. in 1979 introduced maximum power point tracking to the scientific world. In order to improve the tracking time and MPPT efficiency, it is followed by a high number of researchers [11, 12]. The main goal of the MPPT method is to get the most power out of the PV source. The researcher's main focus is to design efficient tracking technology that maximizes energy yield at any atmospheric condition by using evolutionary and soft computing techniques. In the literature, the following are the most important parts of a good MPPT technique:

- Accurate and faster tracking of maximum power point.
- High steady state and dynamic efficiency.
- Zero power oscillations during transient and steady state.
- Lower complexity of design and implementation.
- Low cost.
- System-independent design.

Even though several MPPT techniques are proposed in the literature, performance optimization is the major challenge in the field of PV MPPT. While designing an efficient MPPT algorithm, the important aspects of the PV system which need to be considered are given in Figure 4.18.

The MPPT control of the single-peak P–V curve requires simple algorithms like the fixed voltage or current method, perturb and observe (P&O) and incremental conductance (INC). However, the presence of multi-peak P–V curves under PSC degrades the effectiveness of these algorithms [9]. The conventional gradient-based techniques are popular for tracking the MPP of single-peak curves, but these may fail to track the GP of multi-peak P–V curves due to their hill-climbing nature, and it may cause a reduction of the output power of up to 40%. Under PSC, the power loss due to tracking the wrong MPP and mismatch losses causes a total loss of 54% [6]. Hence, there is a need to develop suitable MPPT algorithms to accurately track the GP of the multi-peak P–V curve under PSC, which ensures maximum energy yield from the PV system under varying atmospheric conditions.

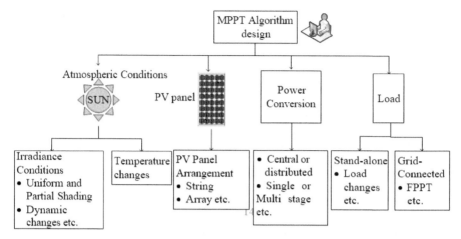

Figure 4.18 MPPT algorithm design considerations.

4.6 Working Principles of MPPT Techniques

MPPT techniques are search methods in the defined search space. These techniques are classified as single and multi-search agent methods based on the number of searching agents used. Single-agent methods mostly use the gradient information of the P–V or I–V curve, which are classified as gradient-based MPPT techniques, whereas multi-search agent methods use more than one searching agent, exploration of the search space is derived from nature. Multi-agent search methods are popular in bio-inspired optimization algorithms, which come under soft computing techniques, because of their system-independent search.

4.6.1 Gradient-based MPPT techniques

Among the conventional gradient-based MPPT techniques, P&O and INC are popular because of their simplicity and ease of implementation. P&O and INC use a single search agent, which is the control variable of the MPPT converter (duty ratio or voltage). Searching MPP starts from the initial value, and the direction of the search is based on the change in power in each perturbation, as shown in Figure 4.19. Perturbations to the control variable are continued even after reaching the MPP, which is required for finding the new MPP for any irradiance changes. However, these continuous perturbations to the control variable, even after reaching MPP, lead to power oscillations in the steady-state tracking. The speed of tracking MPP depends upon the

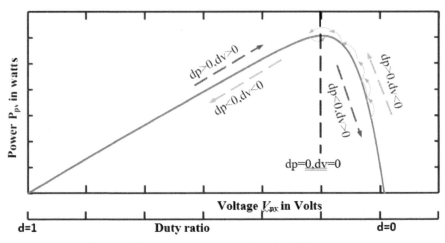

Figure 4.19   Working principle of P&O MPPT technique.

perturbation step size. If a larger perturbation step size is chosen, it leads to faster tracking and high-power oscillations in the steady state, whereas a smaller perturbation step size results in slower tracking and low power oscillations in the steady state. Hence, proper step size is required to balance the tracking time and power oscillations in the steady state [13]. Under irradiance changes, P&O is able to track the first immediate peak from the initial position of tracking, and it may fail to track the GP of the multi-peak P–V curve under PSC due to the hill-climbing nature, as shown in Figure 4.20.

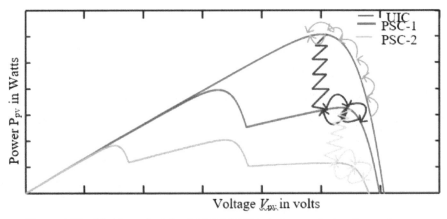

Figure 4.20   Working principle of P&O MPPT technique under irradiance changes.

Therefore, the gradient-based MPPT techniques are best suited for single-peak P–V curves, that is, under uniform irradiance, and these may fail to track GP under PSC.

4.6.2 Soft computing-based MPPT techniques

The soft computing (SC)-based MPPT techniques are multi-search agent methods. The process of exploring the search space is known as exploration, and finding the optimum after satisfying the convergence criteria is known as exploitation. In the bio-inspired SC technique, exploration and exploitation are inherited from either animal's or bird's natural searching strategies. In a well-balanced optimization algorithm, both the exploration and exploitation processes should be strong. When these SC techniques are applied in the field of MPPT to optimise the nonlinear multi-peak P–V or single-peak P–V curve, the objective function is to maximise the PV power. Therefore, the problem formulation for the PV MPPT is given in eqn. (4.5). The constraint in the problem formulation is the control variable limits of the MPPT controller:

$$Maximize \quad P_{pv}(d) \tag{4.5}$$

$$Subject \quad to \quad d_{min} < d < d_{max} \tag{4.6}$$

For example, when meta-heuristic techniques like particle swarm optimization (PSO) are applied to the online PV MPPT problem, the exploration of the search space is explained as follows [14]. First, the exploration of the search space starts with the selection of the number of search agents (particles) and the initialization of the particles (duty ratios) in the search space. For better understanding, the exploration phase of PSO in one iteration is explained as shown in Figure 4.21. The current position of the particle X^k = 0.6 duty ratio changes its position to X^{k+1} = 0.5 based on the eqn. (7) which is derived from birds flocking nature in the swarm. The current position of the particle moves more toward the $G_{best}i$ = 0.35. Therefore, the next position of the next particle is X^{k+1} = 0.5 with the assumed velocity vel^{k+1} = 0.1, which can be calculated with eqn. (8):

$$X_i^{k+1} = X_i^k + vel_i^{k+1} \tag{4.7}$$

$$vel_i^{k+1} = W \times vel_i^k + c_1 \times r_1 \times (P_{best i} - X_i^k) + c_2 \times r_2 \times (G_{best i} - X_i^k) \tag{4.8}$$

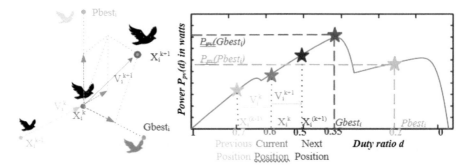

Figure 4.21   Working principle of PV MPPT technique based on PSO.

where i indicates particle number, k represents the current iteration, w indicates weight factor ranges from 0.4 to 0.9, c_1 indicates a cognitive factor ranges from 0.1 to 2, c_2 represents social factor ranges from 0.1 to 2, and r_1 and r_2 are random numbers $\in R\ (0,1)$. In order to move the particle position more toward the global best, $c_2 > c_1$ is chosen in the velocity update equation.

4.7 Classification of MPPT Techniques

In recent years, the main focus of researchers in the field of PV systems has been on extracting maximum available power during all dynamic weather conditions like PSC and optimal performance of the tracking technique. In the literature, several MPPT techniques are proposed to operate the PV system at the maximum power point on the nonlinear P–V curve, which include gradient and advanced SC-based MPPT techniques [15–17], as shown in Figure 4.22. Further, these MPPT techniques are classified as direct and indirect techniques based on the PV parameters considered for the design of the algorithm. These techniques are further broken down into online and offline PV MPPT techniques based on the data and sensors needed to run the algorithm.

Under varying weather conditions, to improve the tracking speed and power oscillations of the conventional gradient-based MPPT techniques (P&O and INC), hybrid methods are implemented in the literature [17, 18], whereas these hybrid methods are created by combining the benefits of traditional gradient and AI-based MPPT techniques. However, because of the complexity of the fuzzy logic controllers and the high cost of implementation because of advanced digital controllers, these hybrid techniques are not popular [19]. The neural networks and fuzzy logic-based AI MPPT

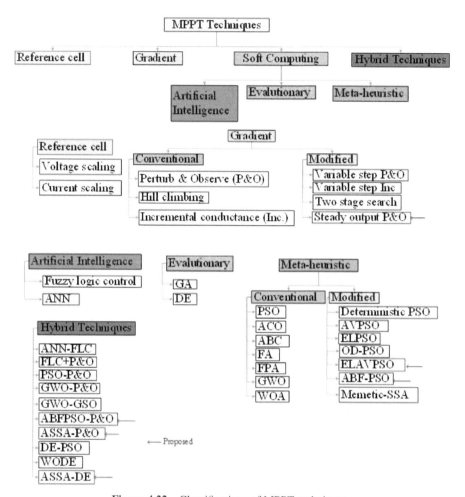

Figure 4.22 Classifications of MPPT techniques.

techniques show rapid convergence under dynamic weather conditions. However, it requires a huge amount of offline training [20]. These techniques are system-dependent and require irradiance, temperature, open-circuit voltage and short-circuit current of each panel in the PV array. Hence, the sensors become more, and the cost of the system increases.

Since finding GP on a multi-peak nonlinear P–V curve within a defined search space is an optimization problem, SC techniques based on swarm intelligence (SI) are used for obtaining optimal solutions by using exploration and exploitation phases. In the literature, swarm-based algorithms like PSO, grey wolf optimisation (GWO), whale optimisation (WO) and evolutionary algorithms like differential evolution (DE) are much explored for PV MPPT. The main advantage of SI techniques is:

- Ability to track GMPP under PSC.
- Zero steady-state oscillations.
- High steady-state efficiency.
- Lower complexity of design and implementation as compared to other AI techniques.
- System-independent design.

However, the major drawbacks are:

- Premature convergence.
- Parameter tuning for varying atmospheric conditions.
- Higher power oscillations during the exploration phase.
- Slow tracking of MPP as compared to conventional P&O and INC.

An adverse effect of the above drawbacks on PV MPPT is high energy loss. Hence, overall PV system conversion efficiency is decreased.

4.7.1 Gradient-based MPPT techniques

Traditional gradient-based MPPT techniques like P&O and INC are popular because of their simplicity and easy implementation. Under UIC, since the P–V curve is a single-peak curve, conventional MPPT algorithms like P&O and incremental conductance track the MPP accurately. However, these techniques' performance is degraded under non-uniform irradiance conditions and large irradiance changes [21]. Another drawback of these methods is that continuous perturbation after reaching the maximum power point causes power and voltage oscillations in steady-state tracking [22]. Hence, to improve the tracking speed and power oscillations of the conventional gradient-based MPPT techniques (P&O and INC), hybrid methods are implemented in the literature [17, 18], whereas these hybrid methods are created by combining the benefits of traditional gradient and AI-based MPPT techniques. The AI techniques are used to find the optimum value of perturbation step size for the dynamic atmospheric conditions. However, these hybrid techniques also fail to track the PSC's multi-peak P–V curve's global peak due to the hill-climbing nature of gradient-based techniques. The failure to track the GP of a multi-peak P–V curve causes power loss in addition to mismatch loss of the PV system, which may cause an overall reduction of the output power of up to 54%. Therefore, the main limitation of the conventional gradient-based MPPT techniques is their failure to track the GP of multi-peak P–V curves, which may cause a reduction of output power and net energy yield. This is mainly due to tracking only the first immediate peak from the starting point of tracking because of its hill-climbing nature. Hence, to avoid the failure to track the GP of the multi-peak P–V curve under PSC and achieve faster and more accurate MPP tracking with low-power oscillations, conventional gradient-based MPPT techniques are combined with SC-based MPPT techniques.

4.7.2 Soft computing-based MPPT techniques

SC techniques can optimise the multiple maxima power curve under PSC and these are also called global maximum power point tracking (GMPPT) techniques [23]. PV model-independent and zero steady-state power and voltage oscillations are the main advantages of SC techniques [14]. However, these techniques suffer from large initial power oscillations during tracking due to the spread of search throughout the power curve in the exploration phase. Improper algorithm control parameters result in premature convergence, causing high energy loss due to the wrong optimum being found. The frequent, continuous and rapid change in irradiance can cause a failure of the GMPPT algorithm in the form of a local optimum. Hence, there is a need to derive modified MPPT techniques, which can reduce the drawbacks like complex parameter tuning, initial power oscillations and MPP tracking failure under dynamic irradiance conditions. Among all the SI techniques, PSO is the most popular in solving PV MPPT problems because of its simple structure and easy implementation. However, the three fixed control parameters of the PSO are not best suited for varying atmospheric conditions. These may degrade the tracking performance of PSO, which may cause premature convergence, that is, failing to track the global peak. Hence, to improve the performance of any soft computing technique, parameter tuning is essential and these techniques are named adaptive techniques. The conventional PSO algorithm needs to be modified to match the requirements of the efficient PV MPP tracking technique. Therefore, the main limitations of SI techniques are large initial power oscillations, premature convergence, improper parameter setting and slow tracking. These limitations result in a low energy yield from the system. To overcome the large initial power oscillations, the velocity of the particles must be controlled and the exploration phase should be limited only to finding the GP region, which can give fast tracking also.

To overcome premature convergence, algorithm parameters need to be tuned properly.

4.7.3 Hybrid MPPT techniques

To improve the performance of standard versions of SI techniques like faster and more accurate tracking of MPP, several modified versions or hybrid MPPT techniques are proposed for PV systems under dynamic weather conditions [24]. Hybrid tracking techniques can be formed by combining two or more conventional or modified SC and gradient-based MPPT techniques. The complexity of the algorithm is increased with hybridisation. However, the essential need to track the global maximum power point under all dynamic weather conditions encourages researchers to investigate more advanced hybrid tracking techniques.

For rapidly varying weather conditions, hybrid MPPT techniques, combining the advantages of artificial intelligence and conventional gradient-based MPPT techniques, are proposed [25, 26]. But, the implementation of these techniques is sophisticated and requires extensive offline training. The new grid codes impose MPPT for all grid-connected systems to get maximum energy yield and active power limit control to maintain stability under peak power production periods. So, the MPPT tracking algorithm should be able to find the maximum power point in all types of changing weather and have active power limit control in the form of FPPT, which is a cost-effective way to control active power.

In stand-alone PV systems, the sudden change in load causes a change in the operating point from the maximum power point. The change in load causes the reinitialisation of the SI technique, which results in large power oscillations due to the exploration phase. This reinitialisation of the SI technique is best suited only for insolation changes where global peak changes, whereas for load change, only the operating point is changing but not the global peak.

4.8 Limitations of Traditional MPPT Techniques

A literature review of the MPPT techniques under partial shading conditions leads to the following motivations.

To mitigate the partial shading effects, connecting a bypass diode across the group of cells in a module is cost-effective. However, the resulting multi-peak P–V curve becomes a limitation for the popular conventional gradient-based MPPT techniques like P&O and Inc. The failure of GP tracking resulted in power loss in addition to mismatch loss. The power loss due to the wrong optimum found by the GMPPT technique results in less energy yield under PSC. Hence, the overall PV system conversion efficiency decreases. The author of this chapter is motivated to conduct additional research on the MPPT techniques of solar PV systems under partial shading conditions and to propose efficient tracking technologies for solar PV systems that provide the highest energy yield under PSC.

Even though PSO-based GMPPT techniques have evolved as the most popular GMPPT techniques among SI-based MPPT techniques, they have drawbacks like complex parameter tuning, initially large power oscillations and premature convergence. Hence, modified PSO-based MPPT techniques are proposed in the literature to get faster and more accurate tracking of GMPP. Under dynamic weather conditions, identifying GMPP with less probability of premature convergence is a major challenge for all GMPPT techniques. The large initial power oscillations due to the exploration of the PSO algorithm should be limited in online PV MPPT. Hence, there is a need for further investigations into PSO-based MPPT techniques to get fast and accurate tracking of GMPP under PSC with fewer power oscillations. To avoid the major drawbacks of PSO: weak exploration and more number of parameters (three) to be tuned, there is a need for a simple GMPPT technique with fewer parameters tuning.

Salp swarm algorithm is a bio-inspired soft computing technique, already proved to be a simple and efficient optimisation algorithm with one control parameter for most of the engineering design problems which also matches with the requirement of fewer control parameters and more efficient tracking of PV MPPT.

4.9 Advanced MPPT Techniques

Advanced GMPPT techniques are proposed to overcome the limitations of state-of-the-art MPPT techniques for PV systems under partially shaded conditions. The main focus of this chapter is to discuss the advanced hybrid MPPT techniques. With these MPPT techniques, the PV system will produce the most energy possible even when it is partially shaded.

(1) ELAVPSO-based GMPPT technique

To overcome the stagnation of search at local maximum and slower convergence in a PSO-based online PV MPPT technique:

(1) A new fast-tracking PSO-based GMPPT technique (enhanced leader adaptive velocity PSO (ELAVPSO)) with adaptive parameter tuning and enhanced leader features is proposed in [27], and the results are shown in Figures 4.23 and 4.24. From the simulation results, it is evident that the P&O and the proposed ELAVPSO MPPT technique result in almost the same results under uniform irradiance. However, under PSC, only the proposed method tracks the GMPP, that is, $P_{mp} = 34.5$ W.

(2) A new shading detection technique to identify the type of shading based on variable step P&O is proposed to limit the GMPPT technique only to PSC. Hence, power oscillations due to the GMPPT technique under uniform irradiance can be avoided. The results of the proposed shading detection are shown in Figure 4.25. From the simulation and experimental results, it is evident that the proposed shading detection accurately identifies the uniform and PSC conditions and allows the use of the ELAVPSO GMPPT technique only under the occurrence of PSC.

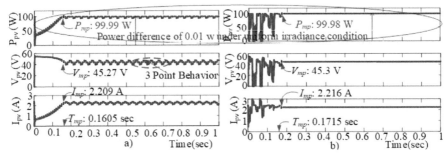

Figure 4.23 MPPT techniques results under uniform irradiance (a) P&O and (b) ELAVPSO.

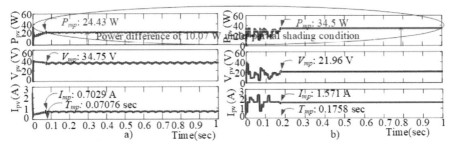

Figure 4.24 MPPT technique results under partial shading condition: (a) P&O and (b) ELAVPSO.

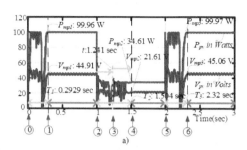
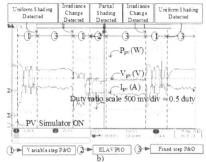

Figure 4.25 Proposed shading detection technique results in (a) simulation and (b) experimental.

In ELAVPSO, mainly premature convergence and parameter tuning are minimised. However, the other drawback of the PSO-based GMPPT technique is that high power oscillations during the exploration phase remain when ELAVPSO is used alone without shading detection.

(2) ABFPSO-P&O-based GMPPT technique

To overcome the limitations of PSO-based MPPT methods: high power oscillations during the exploration phase, stagnation of search at local maximum and slower convergence in a PSO-based online PV MPPT technique:

(1) A hybrid global MPP tracking method based on butterfly PSO (BF-PSO) and P&O algorithms are proposed for a PV system under partially shaded conditions [28]. The results are shown in Figure 4.26. From the results obtained, it is evident that the proposed GMPPT technique results in global MPP tracking in less than 2.4 sec.

(2) A new reinitialisation method based on the previous history of tracking global peak regions under both medium and severe nonuniformity in irradiance changes is proposed. From the experimental results shown in Figure 4.26, it is evident that the proposed re-initialisation method results in faster and more accurate tracking, that is, $T_{mpp} = 1.6$ sec and $T_{mpp} = 0.8$ sec under irradiance changes.

In adaptive BF-PSO, mainly premature convergence, parameter tuning and power oscillations during tracking are minimised. Further, the reinitialisation of particles under irradiance changes improves the tracking speed as well as limits the power oscillations during tracking. However,

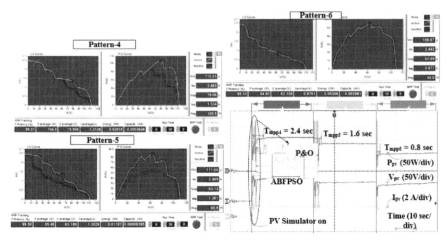

Figure 4.26 Experimentally obtained results of the proposed ABFPSO-P&O MPPT technique.

there are more parameters, that is, three to be tuned in PSO-based MPPT techniques.

(3) ASSA-P&O-based global MPP tracking technique

The GMPP tracking technique based on the adaptive salp swarm algorithm and P&O technique is proposed for a PV string under partially shaded conditions [29], which reduces parameter tuning complexity. It is evident that the proposed global MPPT technique results in faster and more accurate tracking of MPP with a single control parameter.

In ASSA-P&O, the number of parameters to be tuned and power oscillations during tracking are minimised. However, slower tracking of GP region identification in hybrid MPPT techniques (formed with SC and gradient MPPT techniques) is mainly due to the least fitness particles. When these techniques are applied to off-grid PV systems, reinitialisation of the GP region identification stage for load changes degrades hybrid MPPT techniques' performance.

(4) ASSADE-P&O-based GMPPT technique

In this control strategy proposed in [30], the least fitness particle is accelerated more toward the leader to get faster GP region identification. The leader salp is enhanced in the food source region to get better accuracy. In an off-grid PV system, reinitialisation of GP region identification only for irradiance change and direct duty ratio calculation for load change is adopted. Hence, in this method,

(1) A hybrid GMPP tracking of the PV system under partially shaded conditions and load variation is proposed using the adaptive salp swarm algorithm and differential evolution – perturb and observe techniques. The proposed hybrid MPPT technique results in faster and more accurate MPP tracking, whereas direct duty ratio calculation helps in identifying GMPP within a short time of 0.1 sec without reinitialising the SI technique given in Figure 4.28.

The ASSADE-P&O-based GMPPT technique with direct duty ratio calculation under load changes has been established as a favorable method

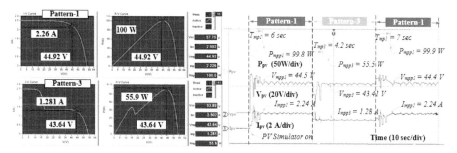

Figure 4.27 Experimentally obtained results of the proposed ASSA-P&O MPPT technique.

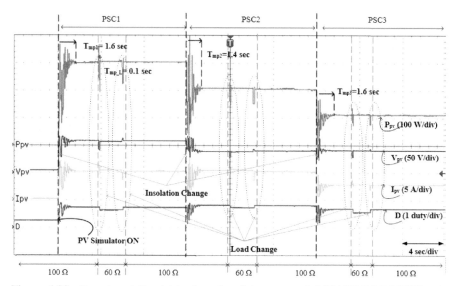

Figure 4.28 Experimentally obtained results of the proposed ASSADE-P&O MPPT technique under load changes.

for the off-grid PV system. However, for a grid-connected PV system, the hybrid tracking technique's performance optimisation must meet the grid standards like active power limit control.

4.10 Conclusion

A comprehensive literature review on the maximum power point tracking techniques of solar PV systems under partial shading conditions has been presented. The shortcomings of all existing approaches have been discussed. Under PSC, conventional gradient-based MPPT techniques such as P&O may fail to track the global MPP of a multipeak P–V curve. Therefore, soft computing-based MPPT techniques were invented. However, conventional soft computing-based GMPPT techniques have shortcomings like algorithm control parameters that are not best suited for all dynamic weather conditions. Hence, hybrid MPPT techniques were invented to get fast and accurate tracking of GMPP under partially shaded conditions.

References

[1] S. K. Gupta and R. S. Anand, "Development of Solar Electricity Supply System in India: An Overview," *Journal of Solar Energy*, pp. 1–10, 2013.

[2] A. Urja, "Newsletter of the Ministry of New and Renewable Energy, Government of India," vol. 12, no. 5, apr 2019.

[3] T. V. Dixit, A. Yadav, and S. Gupta, "Experimental assessment of maximum power extraction from solar panel with different converter topologies," *International Transactions on Electrical Energy Systems*, vol. 29, no. 2, p. e2712, 2019.

[4] H. Patel and V. Agarwal, "Maximum power point tracking scheme for pv systems operating under partially shaded conditions," *IEEE Transactions on Industrial Electronics*, vol. 55, no. 4, pp. 1689–1698, April 2008.

[5] M. Balat, "Solar technological progress and use of solar energy in the world," *Energy Sources, Part A: Recovery, Utilization, and Environmental Effects*, vol. 28, no. 10, pp. 979–994, 2006.

[6] A. Mäki and S. Valkealahti, "Power losses in long string and parallel-connected short strings of series-connected silicon- based photovoltaic modules due to partial shading conditions," *IEEE Transactions on Energy Conversion*, vol. 27, no. 1, pp. 173–183, 2012.

[7] J. Cubas, S. Pindado, and C. D. Manuel, "Explicit Expressions for Solar Panel Equivalent Circuit Parameters Based on Analytical Formulation and the Lambert W-Function," *Energies*, vol. 7, no. 7, pp. 1–18, June 2014.

[8] E. A. Sweelem, F. H. Fahmy, M. M. A.-E. Aziz, P. Zacharias, and A. Mahmoudi, "Increased efficiency in the conversion of solar energy to electric power," *Energy Sources*, vol. 21, no. 5, pp. 367–377, 1999.

[9] A. Dadjé, N. Djongyang, J. D. Kana, and R. Tchinda, "Maximum power point tracking methods for photovoltaic sys- tems operating under partially shaded or rapidly variable insolation conditions: a review paper," *International Journal of Sustainable Engineering*, vol. 9, no. 4, pp. 224–239, 2016.

[10] H. D. Tafti, A. I. Maswood, G. Konstantinou, J. Pou, and F. Blaabjerg, "A general constant power generation algorithm for photovoltaic systems," *IEEE Transactions on Power Electronics*, vol. 33, no. 5, pp. 4088–4101, 2018.

[11] B. Subudhi and R. Pradhan, "A comparative study on maximum power point tracking techniques for photovoltaic power systems," *IEEE Transactions on Sustainable Energy*, vol. 4, no. 1, pp. 89–98, Jan 2013.

[12] M. Abdel-Salam, M. T. EL-Mohandes, and M. Goda, "History of maximum power point tracking," *Green Energy and Technology*, pp. 1–29, 2020.

[13] N. Femia, G. Petrone, G. Spagnuolo, and M. Vitelli, "Optimization of perturb and observe maximum power point tracking method," *IEEE Transactions on Power Electronics*, vol. 20, no. 4, pp. 963–973, July 2005.

[14] Y. Liu, S. Huang, J. Huang, and W. Liang, "A particle swarm optimization-based maximum power point tracking algorithm for pv systems operating under partially shaded conditions," *IEEE Transactions on Energy Conversion*, vol. 27, no. 4, pp. 1027–1035, Dec 2012.

[15] M. F. N. Tajuddin, M. S. Arif, S. M. Ayob, and Z. Salam, "Erratum to the 'perturbative methods for maximum power point tracking (mppt) of photovoltaic (pv) systems: a review' international journal of energy research 2015; 39:1153–1178," *International Journal of Energy Research*, vol. 39, no. 12, pp. 1720–1720, 2015.

[16] R. Rawat and S. S. Chandel, "Review of maximum-power-point tracking techniques for solar-photovoltaic systems,"*Energy Technology*, vol. 1, no. 8, pp. 438–448, 2013.

[17] M. Seyedmahmoudian, B. Horan, T. K. Soon, R. Rahmani, A. M. Than Oo, S. Mekhilef, and A. Stojcevski, "State of the art artificial intelligence-based mppt techniques for mitigating partial shading effects on pv systems – a review," *Renewable and Sustainable Energy Reviews*, vol. 64, no. C, pp. 435–455, 2016.

[18] Sharma, A., Sharma, A., Jately, V., Averbukh, M., Rajput, S., & Azzopardi, B. (2022). A Novel TSA-PSO Based Hybrid Algorithm for GMPP Tracking under Partial Shading Conditions. Energies, 15(9), 3164.M. Seyedmahmoudian, T. K. Soon, B. Horan, A. Ghandhari, S. Mekhilef, and A. Stojcevski, "New armo-based mppt technique to minimize tracking time and fluctuation at output of pv systems under rapidly changing shading conditions," *IEEE Transactions on Industrial Informatics*, pp. 1–1, 2019.

[19] A. Tavakoli and M. Forouzanfar, "A self-constructing lyapunov neural network controller to track global maximum power point in pv systems," *International Transactions on Electrical Energy Systems*, vol. n/a, no. n/a, p. e12391.

[20] E. Koutroulis and F. Blaabjerg, "Overview of maximum power point tracking techniques for photovoltaic energy production systems," *Electric Power Components and Systems*, vol. 43, no. 12, pp. 1329–1351, 2015.

[21] S. Bhattacharyya, D. S. K. Patnam, S. Samanta, and S. Mishra, "Steady output and fast tracking mppt (soft mppt) for p o and inc algorithms," *IEEE Transactions on Sustainable Energy*, pp. 1–1, 2020.

[22] Singh A, Sharma A, Rajput S, Bose A, Hu X. An Investigation on Hybrid Particle Swarm Optimization Algorithms for Parameter Optimization of PV Cells. Electronics. 2022; 11(6):909.

[23] K. Aygül, M. Cikan, T. Demirdelen, and M. Tumay, "Butterfly optimization algorithm based maximum power point tracking of photovoltaic systems under partial shading condition," *Energy Sources, Part A: Recovery, Utilization, and Environmental Effects*, vol. 0, no. 0, pp. 1–19, 2019.

[24] S. Rajput, E. Bender, M. Averbukh, Simplified algorithm for assessment equivalent circuit parameters of induction motors, IET Electric Power Applications, 14 (2020) 426-432.M. Chamanpira, M. Ghahremani, S. Dadfar, M. Khaksar, A. Rezvani, and K. Wakil, "A novel mppt technique to increase accuracy in photovoltaic systems under variable atmospheric conditions using fuzzy gain scheduling," *Energy Sources,*

Part A: Recovery, Utilization, and Environmental Effects, vol. 0, no. 0, pp. 1–23, 2019.

[25] Sharma, A., Sharma, A., Averbukh, M., Rajput, S., Jately, V., Choudhury, S., & Azzopardi, B. (2022). Improved moth flame optimization algorithm based on opposition-based learning and Lévy flight distribution for parameter estimation of solar module. Energy Reports, 8, 6576-6592.

[26] D. K. Mathi and R. Chinthamalla, " A hybrid global maximum power point tracking method based on butterfly particle swarm optimization and perturb and observe algorithms for a photovoltaic system under partially shaded conditions.," in Wiley International Transactions on Electrical Energy Systems, vol. 30, no. 10, e12543, 2020, doi.org/10.1002/2050-7038. 12543.

[27] D. K. Mathi and R. Chinthamalla, " Global maximum power point tracking technique based on adaptive salp swarm algorithm and P&O techniques for a PV string under partially shaded conditions.," in Taylor & Francis; Energy Sources, Part A: Recovery, Utilization, and Environmental Effects, 2020, doi: 10.1080/15567036.2020.1755391.

[28] D. K. Mathi and R. Chinthamalla, " A Hybrid Global Maximum Power Point Tracking of Partially Shaded PV System under Load Variation by Using Adaptive Salp Swarm and Differential Evolution – Perturb & Observe Technique.," in Taylor & Francis; Energy Sources, Part A: Recovery, Utilization, and Environmental Effects, 2020, doi:10.1080/15567036.2020.1850927.

5

Application of a Polymer Nanocomposite for Energy Harvesting

Sonika[1], Sushil Kumar Verma[2], Sunil Dutt[3], Vishwanath Jadhav[4], and Gopikishan Sabavath[5]

[1]Department of Physics, Rajiv Gandhi University, Doimukh, Itanagar, India
[2]Department of Chemical Engineering,
Indian Institute of Technology Guwahati, India
[3]Department of Chemistry, Government Post Graduate College, India
[4]Department of R&D (NPD), Deep Plast Industries, India
[5]Faculty of Science, Department of Physics, University of Allahabad, India
E-mail: sonika.gupta@rgu.ac.in; sushilnano@gmail.com;
sunildutt.iitmandi@gmail.com; vjadhav11096@gmail.com;
gopikishan@allduniv.ac.in

Abstract

Recent years have experienced a significant rise in research in advanced polymeric materials as promising materials for the development of the next generation of energy harvesting devices. For lightweight and low-power electronic devices, they have been investigated by researchers for more than two decades as an alternative to standard power sources (such as batteries). A continuing problem with portable, remote and implantable devices is the limited duration of batteries and the demand for regular recharging or replacement. Thermal, vibrational and solar energy are the three major sources of ambient energy. Organic electronic devices such as light photovoltaics and flexible displays have arisen from the development of conjugated polymers and derivative-related processing techniques. These innovations have recently been exploited to produce organic thermoelectric materials, which have the potential to be utilised in wearable heating and cooling devices as well as energy generation at temperatures close to ambient temperature. The highest thermoelectric materials so far have been inorganic compounds and multiwall carbon nanotubes, which are generated by immensely difficult chemical synthesis (in situ and ex situ) processing routes and they have relatively low earth abundance. In addition to showing the figures of merit that are equivalent to those of these inorganic materials and MWCNT, molecular materials and hybrid organic–inorganic materials also exhibit novel transport behaviors that point to previously conceivable optimisation routes and device designs. Fundamentally, a thermally non-conductive but electrically linked nanostructured network should be developed for conjugated polymer nanocomposite-based thermoplastic (TP) materials.

Keywords: Conjugated Polymer and Derivative; Multiwalled Carbon Nanotube; Inorganic Materials; Thermoelectric; Seebeck Coefficient.

5.1 Introduction

Fossil fuels were among those examined because they have been the world's main energy source since the 19th century, and the growing world's population is rushing toward them without attaining the potential adverse effects they could have on the environment. Energy is the major part from the start to the finish of a day directly or indirectly in forms of thermal, electrical, chemical, nuclear, or/and many other.

In November 2018, with the assistance of the EU, decisions between the Council of Ministers, the European Parliament and the European Commission were integrated to design the most modern requisite objectives as taking into account the future prospects. The proposed standards suggest a minimum of 40% reduction in greenhouse gas emissions from 1990 levels. Additional incentives have been offered in the form of a minimum of 32% growth in renewable energy and a minimum of 32.5% rise in efficiency with reference to the 2007 targets. Collective efforts must be made to reduce carbon emissions and obtain renewable energy in order to achieve the proposed plans, which will reduce emissions to the environment [2].

In sub-Saharan Africa, Alemán-Nava et al. [3] explored that inclusive development and sustainable energy influence CO_2 emissions. The findings imply that CO_2 emissions are notably adversely impacted by renewable energy. This signifies that raising renewable energy sources will help to reduce CO_2 emissions in the region. Moreover, they noted that the use of renewable energy improves environmental sustainability in a region where inclusive development is still lagging and adding to CO_2 emissions. Similarly, CO_2 emissions from the use of liquid fuel were re-estimated using the informative variables and the findings continued to be consistent. Biomass, hydropower, geothermal, wind, solar and other naturally occurring sources with consistent or limited supply are illustrations of renewable energy sources [3]. Uranium, petroleum, natural gas, coal and other non-renewable energy sources include others. Energy harvesting integrates surplus energy from all available energy sources (renewable or non-renewable).

Energy harvesting technology typically operates low-power electronic devices by converting waste energy into valuable electricity [4]. The operation of the thermoelectric generator (TEG) has been mostly liable for energy harvesting in thermoelectric forms [5]. Using semiconductors that express the photovoltaic effect and piezoelectric technology to reduce low-frequency vibration, acoustic noise, human motion, etc., photovoltaic renewable energy transforms solar radiation (from both indoor and outdoor resources) into direct current electricity [6].

5.2 Types of Energy Harvesting Systems

5.2.1 Thermoelectric

Thermoacoustic is the conversion of a temperature difference into an electric current and vice versa. This reverse energy conversion, from thermal to electric and vice versa, makes thermoelectric generator special. All energy applications in thermoelectric are based on the concepts of Seebeck and Peltier effects. Because charge carriers in semiconductors and metals are able to move as gas molecules, while conveying charge and heat, thermal effects occur [6]. Because they must have different electron densities, the thermo-electric module (TEM) is made of two unique semiconductor materials, most commonly bismuth telluride (Bi_2Te_3) (one n-type and one p-type) [7].

The Seebeck model implies that a TEG transforms heat directly into electrical energy. The mechanism with which thermal energy generates electrical energy is known as the Seebeck effect. The Seebeck effect is achieved when a temperature difference across a conductor results in a voltage at the ends of the conductor. The Peltier effect is the technique by which electrical energy is transformed into thermal energy. In these situations, the thermoelectric cooler (TEC) is the tool of choice [8].

The productivity of a thermoelectric generator is key for the capture of thermal energy (TEG). Thermoelectric generators are static, immobile devices. The major challenge in thermoelectric energy harvesting is the imposition of a wide temperature gradient. It has features such as silent, scalable and reliable, making it ideal for tiny diffused power generation. Due to the small surface heat exchanges and high thermal conductivity of thermoelectric materials, this requires enormous heat fluxes [9]. Numerous TEG applications have been described by Zheng et al. [10] in a diversity of multifunctional industries, including the automotive, aerospace, industrial, household and thin-film sectors. TEG employs waste thermal energy from the automotive, aerospace, industrial, etc. domains to produce electrical energy.

The TE refrigerator/cooler can be exploited for therapeutic diagnostics such as the retention and transportation of vaccines, blood serum and biological substances, as well as various operations, according to study [10–12]. Most worldwide manufacturers have developed their own exhaust heat recovery systems utilising TEG, such as Renault, Honda, Ford and others [13–15]. Papkin et al. [16] proposed TEG, which produces electricity using the waste heat of a car's engine cooling system and they arrived at the conclusion that TEG reduces fuel consumption by 3–5% and produces 500–700 watts of electrical power using 4–6% of the cooling system's thermal energy. TE units are utilised in heating and cooling automotive seats by several automakers, include Hyundai, Jaguar, Range Rover, Toyota and GM [17].

5.2.2 Piezoelectric energy harvesting

Through piezoelectric effect, mechanical stress is transferred into electrical potential. This strain comes from a wide range of places, examples include movements of people, low-frequency seismic vibrations and sound. Walking can produce mechanical energy using piezoelectric effect [18].

Assured crystalline materials, such as quartz, Rochelle salt, tourmaline, topaz and barium titanate ($BaTiO_3$), can naturally generate energy when pressure is applied, piezoelectricity, or pressure electricity, can be observed. A sensor or energy transducer can be applied with this, which is known as the direct effect. The reverse effect, on the other hand, pertains to the deformation that these crystals experience in the presence of an electric field and can be utilised as an actuator. In addition, several biological materials, such silk [19], wood [20], bone [21], hair [22], rubber [23] and dentin, exhibit the piezoelectric effect. A high energy conversion rate

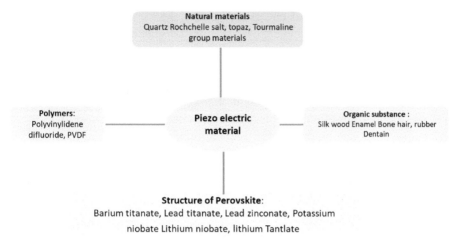

Figure 5.1 Different types of piezoelectric materials.

is demonstrated for synthetic materials with significant electromechanical coupling constants, such as ceramics made of barium titanate [24], potassium niobate [25], sodium tungstate [26] and lead zirconate titanate (PZT) [27]. In contrast, piezopolymers are significantly more flexible than piezoceramics but have smaller electromechanical coupling constants. For sensor applications, a material with a better piezoelectric coefficient, such as polyvinylidene difluoride, PVDF or PVF2 [28, 29] is used.

When converted into valuable electrical energy, this energy can be used to power wearable electronic devices including sensors and GPS receivers. Piezoelectric sensors, in particular those with high-frequency sound, are used in ultrasound transducers for medical imaging and industrial non-destructive testing [18]. Several bending beam systems have observed the stress produced for each type in specific geometrical limitations and material properties and have established the use of piezo films effective for shoe inserts and walking-like excitation [30]. The appropriate design can be determined by studying the electricity that has been harvested. Piezoelectric membranes were developed within the shoe by Granstrom et al. [31] mainly on the basis of their preliminary study. A brand-new backpack was developed with a PVDF substrate for energy harvesting that can generate power from the forces that are different between the users and the backpack. They also considered evaluating the performance of the electromechanical conversion qualities between a PVDF and an ionic conductive polymer transducer for energy harvesting. Hansen et al. [32] developed a hybrid energy harvesting device that harvests both biomechanical and biological energy using a PVDF piezoelectric nanofiber generator. They observed that the two different methods of energy harvesting would be used simultaneously or separately to increase output and durability.

With an emphasis on the development of coaxial nanofibers as conversion components from high-efficiency piezoelectric material PVDF, Liu et al. [33] introduced a small energy device for medical microrobots that are intended to function in blood vessels. Using polyurea thin-film vibration, Koyama et al. [34] observed the development of electric power. Using finite element analysis of cantilever generation, the capacity to transform mechanical energy to electrical energy was evaluated. A thinner, shorter cantilever configuration with a higher resonance frequency was used to enhance conversion efficiency. Tien et al. [35] examined piezocomposite multi-layered energy harvesting materials made of glass/epoxy, PZT ceramic and carbon/epoxy. After numerical and experimental validation, it is asserted that piezocomposite has the ability to capture vibrational energy.

5.2.3 Pyroelectric energy harvesting

Since some materials exhibit the pyroelectric property, which leads them to briefly generate an electrical voltage when heated or cooled, photovoltaic materials have been used to convert thermal energy into electrical energy.

Lepidolite, a mineral and synthetic materials to pyroelectric properties, such as gallium nitride, a semiconductor, cobalt phthalocyanine, caesium nitrate and lithium tantalate, a crystal of piezoelectric properties were the first materials to exhibit the above property and have been used to generate pyroelectric fusion on a small scale [37, 38]. Triglycine sulfide, doped hafnium oxide and polymers like polyvinylidene fluoride (PVDF), which have recently been revealed to have pyroelectric and piezoelectric properties and exhibit low heat conductivity and low permittivity, are all exploited as sensors in devices like burglar alarms.

On a basis of PVDF thin film, Zhang et al. [40] display pyroelectric nanogenerators' scalability. The output short-circuit current and open-circuit voltage were 0.8 A and 8.2 V, respectively. With a 0.1 mW load, the maximum output power was measured to be 2.2 mW. An LCD or LED display can be sustained by the energy generated, or it can be retained in a capacitor for later application. Similar to the above, Yang et al. [41] also exhibited a flexible hybrid energy harvesting cell composed of nanothermal, piezoelectric nanogenerators and a solar cell, which can harvest thermal, mechanical and solar energies separately or simultaneously.

5.3 Polymer-based Energy Harvesting System

The solar, thermal, radio frequency and piezoelectric energy harvesting technologies are the most frequently used. Light energy is transformed into electricity employing photovoltaic (PV) or solar cells. Within all energy collectors, photovoltaic cells have the highest energy density and output power.

Energy harvesting is the technique of capturing and converting minute amounts of easily available energy from the environment into useful electrical energy. Instead, electrical energy is produced for instant use or it is captured and stored for future use. This supplies a substitute energy source for use in instances when there is no grid power and the installation of wind turbines or solar panels is inefficient. Small power sources, in comparison to external solar electricity, do not provide a significant amount of power. However, in the lower portions of the power spectrum, the collected power is suitable for the entirety of wireless, remote sensing, body implants, RFID and other applications. Moreover, even if the energy generated is adequate to power the device, it can still be used to extend the battery life. Energy harvesting also comes by the terms small energy harvesting and energy harvesting. Because it presents an alternative power source for electronic devices in instances where there are no conventional energy sources, energy harvesting is essential. It has the same benefits for applications in remote areas, underwater and other challenging environments, where conventional batteries and energy are adequate.

Batteries are mainly used to power low-power electronic devices such as integrated and remote sensors. Even long-lasting batteries, however, have a certain survival rate and should be replaced every few years. When there are several sensors dispersed across far-off sites, replacements become expensive. On the other hand, energy harvesting techniques give infinite operational life for low-power equipment and do away with the need to change batteries in instances where doing so is expensive, impractical, or dangerous. Most of the energy harvesting applications are intended to be cost-effective, self-sufficient and just need little to no operation over a long time frame. In addition, power is used closer to the source, removing the necessity for long cables and transmission losses. The software or device can operate without a battery if the power is sufficient to supply it directly.

5.3.1 Plastic solar cell

Conjugated polymers have been studied as light absorbers, electron transfer, acceptors and/or hole-transporting materials in polymer solar cells, also described as plastic solar cells, for the past 20 years [42]. Currently, 10% is the verified maximum energy conversion efficiency [43]. Innovative polymers with diverse molecular architectures and their usage in solar devices are being widely researched in order to improve the concert.

The construction of the polymer solar cell is comparable to a typical silicon-based solar cell with a two-dimensional junction at its center. The P–N junction solar cell capability of this device is predicted by society, and it will only involve a pattern of organic semiconductor materials in the p- and n-type. The produced excitons must be near enough to the junction interface to produce free charge carriers due to the short diffusion length of excitons, which causes the length of those excitons to revert to the ground state. On the other hand, the sunlight-absorbent material is enough thick to absorb as much of the incident sunlight. Donor (p-type) and acceptor (n-type) diffuse phases were successfully developed to attain high performance in order to balance this challenge. A primary polymer solar cell with a high-energy change of the strength of 2.9% under the radiation of 20 mW cm2 was achieved in 1995 employing poly(2-methoxy,5-(2/-ethylhoxy)-1,4-phenylphenylene) (MEH-PPV) and its derivatives [44].

Since the previous generation of solar cells, which predominated for decades commencing in the 1950s, single crystalline silicon solar cells have been classified. Modern materials such as CIGS and CdTe allowed the development of thin-film solar cells, which are now usable as the second generation of solar cells on the market. The materials CIGS and CdTe are expensive and unfavorable on a global level because they contain toxic elements including indium (In), gallium (Ga) and the metal cadmium (Cd), as well as rare elements. The current generation of solar cells includes conjugated organic materials (polymers derivatives, nanoparticles, dyes, etc.), as well as CZTS and perovskite thin films, which are still being studied to enhance performance and stability and so have not yet been used widely in industry. Developments in the manufacturing of cells, new technologies and the production of new, reasonably priced semiconducting materials have resulted in a significant decline in solar display costs, which has resulted in broad acceptance among the many consumers. Moreover, the globally known results of conjugated polymers (CPs) focused on devices in transistors, photovoltaics, photodiodes, light-emitting diodes, etc. These polymers' potential to be solution formed into thin films, which facilitates the development and also lowers material costs, such as device needs, is another major benefit above inorganic analog silicon (Si). A low-bandgap rate can occasionally be generated with an extensive absorption spectrum and a high absorption coefficient. The conducting polymer with allied conjugated donor and acceptor blocks could be produced using block copolymerisation method. In addition, conjugated polymer or its copolymer-based donor is intended to be combined, composited and hybridised with the relevant acceptor. The conjugated polymer not only affirms its superiority as a photovoltaic material applicator, but it also shows significant potential for other advancements to the confidence of overstatement, effective, reasonable, printable solar cells for techno-commercial consequences.

5.3.2 Polymer as light-harvesting material

Power adaptation potency, which is determined by dividing the maximum harvest power with instance power, is used to evaluate the photovoltaic performances of solar cells [45]:

$$\eta = \frac{P_{\max}}{P_{in}} = \frac{J_{SC} \times V_{OC} \times FF}{P_{in}}. \quad (5.1)$$

Here, J_{SC} refers to short-circuit power density, V_{OC} for open-circuit voltage (OCV) and FF for fill factor, which is defined by the subsequent equation and implies a correspondence to the quadrilateral shape of the J–V curve:

$$FF = \frac{J_{P_{\max}} \times V_{P_{\max}}}{J_{SC} \times V_{OC}}. \quad (5.2)$$

It could be determined that the fill issue, short-circuit thickness and open-circuit voltage should be enhanced in order to improve the photovoltaic exhibitions. For the most part, the short-circuit current density straightforwardly connects with the bandgap of light-harvesting materials. Photons having energy from the bandgap can be absorbed and then cause excitons when a solar cell is stimulated. The relation between the obtained photo-generated charges and furthermore the width of incident photons, referred to as external quantum productivity (EQE), is being used to determine the efficiency of the light-harvesting device. The measure of short-circuit current density could represent a condition

$$J_{SC} = e \int_{F_g}^{\infty} EQE_{(E)} \times nAM1.5(E)dE. \quad (5.3)$$

Currently, E_g is the bandgap of light shielding and $nAM1.5(E)$ is the number of photons with totally distinct energies in the regular period range. A reduced bandgap is able to retain more light and so have a higher J_{SC}.

Open-circuit voltage, which is also constant due to the bandgap of photoactive materials, is a constituent of moving the solar cell panel. In polymer solar cells, the open-circuit voltage is primarily associated with energy divergence between the lowest unoccupied molecular orbital (LUMO) of the acceptor and the highest occupied molecular orbital (HOMO) of the donor [46]. By trial and inaccuracy, the following equation describes the functional open-circuit voltage:

$$V_{OC} = \frac{1}{e} \times (|HOMO_{Donor}| - |LUMO_{Acceptor}|) - 0.3\,V. \qquad (5.4)$$

Moreover, because electrons must transit from the source to the acceptor, the LUMO of the acceptor should be situated less than in the donor. Developing with photoactive materials should produce a harmony between the elevated HOMO/LUMO energy discrepancy and the constrained band gap to concentrate on light harvesting value in while also conserving a high open-circuit voltage.

The fill factor, which transfers to the microscale structure of the dynamic layer, charge evacuation layer and post-conduct of admission technique, is the third perspective that influences how a solar-based cell is represented [47]. The short-circuit current thickness and open-circuit voltage, which are comparable to inside series and equal opposition during a circuit, will also be modified by the fill factor.

Remarkable advancements in materials research and innovation have developed efficient strategies for recognising a strong photovoltaic display. Morphology is equally as important in mass heterojunction polymer solar cells as it is in photoactive polymer materials that get these attributes. In designing extremely effective polymer solar cells, it is essential to consider three key points of contention: the materials strategies, the morphology organisation and also the point of interaction.

5.4 Types of Polymers Used for Energy Harvesting System

When using a variety of reaction techniques, such as emulsion, suspension, arrangement and bulk, vinylidene difluoride (VDF) is synthesised into polyvinylidene fluoride, an especially non-thermoplastic fluoropolymer. Then, copolymers of VDF with ethylene and halogenated ethylene monomers

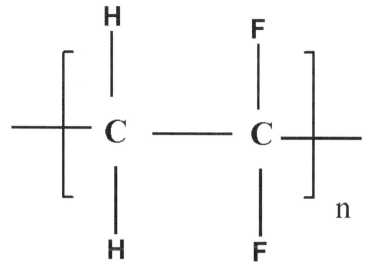

Figure 5.2 Structure of polyvinylidene fluoride (PVDF).

were further developed [48]. These copolymers have a density of 1.78 and showed high attributes comprising resistance to solvents, acids and hydrocarbons. They are also lead-free, reasonable and simple to process.

It is generally used in the medical, healthcare, semiconductor and defense industries, impartial as in lithium-ion batteries, and has a potential to be injected, moulded, or welded. When exposed to light effect and moderate breeze speeds, polymer-based piezoelectric materials yield better maximum voltage/power than fired piezoelectric materials, according to research by Vatansever et al. [49] on energy rally employing PVDF (Figure 4.1) and lead zirconate titanate and graphite. This study has shown that employing adaptable polymer-based piezoelectric designs, it is possible to produce energy from inexhaustible hotspots used in low-power electronic devices in outdoor applications. Zabek et al. [50] developed graphene-ink laminate structures on polyvinylidene fluoride for pyroelectric thermal energy harvesting and waste heat recovery and observed the closed-circuit pyroelectric current for the system of graphene-ink/PVDF/aluminum nanocomposite.

PANI is a specific class of directive polymers that were produced through the polymerisation of aniline, and it can be produced using a selection of novel techniques such as plasma, electrochemistry, format, enzymology and other unique approaches. Again it is classified into heterophase, technique, interfacial, promoting, metathesis, self-collecting and sonochemical

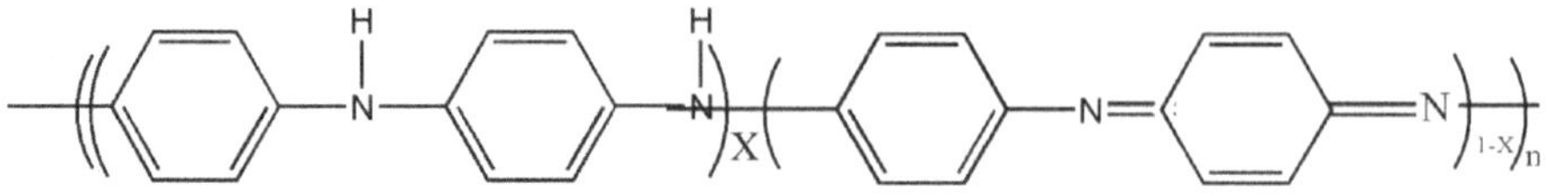

Figure 5.3   Structure of polyaniline (PANI).

polymerisations. PANI (Figure 4.2) is recognised as the most potential due to its inherent qualities, such as its convenience of fusion, low-cost monomer, tuneable properties and better stability compared to other intrinsically conducting polymers, as well as its exquisite biocompatibility and fundamental strength [51]. Electrical conductivity has also been studied extensively in microbial power module as the electrode modifier. The limitations of PANI's potential uses included the infusibility, particularly low solubility in a scale of the various solvents, hygroscopicity and notably poorer conductivity comparable to metals [52].

The ultrahigh dielectric stability that PANI exhibits make it valuable as a capacitor and an energy storage device [53, 54]. As CNT is also a promising material for the development of energy-storing technologies, the composite of CNT/PANI displays improved energy-storing potential.

Zhao et al. [57] and Gambari et al. [58] had designed battery-powered electronics employing PANI, whereas Stakhira et al. [55] and Liu et al. [56] had developed for photovoltaic cells (solar cells) and diodes. Polyaniline was applied to produce heterojunction solar cells by Wang et al. [59] on translucent silicon. Khalifa M et al. [60] developed a highly adaptable electrospun PVDF/PANI/graphitic carbon nitride nanosheets blend nanocomposite (PPBF) nanogenerator for piezoelectric energy harvesting and assumed that it displays better performance as well as proposes power, environmental, adaptability, accessibility and versatility to large-scale manufacturing, making it an extremely encouraging piezoelectric material for energy collecting applications.

By oxidatively polymerising pyrrole monomer, polypyrene (PPy) is a naturally occurring polymer blend that can be synthesised using a variety of methods, such as the Knorr, Paal–Knorr and Hantzsch pyrrole synthesis. Zav'yalov pyrrole polymerisation, Barton–Zard pyrrole incorporation and related reactions, among several other applications [62] PPy (Figure 4.3) is a naturally superior polymer that is employed in the hardware, optical, natural and clinical industries because of its inherent qualities such high environmental strength, slightly easier handling and basic doping-based electrical characteristic modifying [63, 64]. According to Xu et al. [65] and

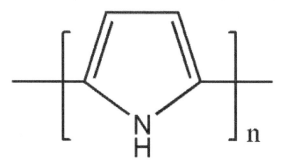

Figure 5.4 Structure of polypyrene (PPy).

Silakhori et al. [66], polypyrene (PPy) has now become to be known as a type of polymer-based natural PCM due to its unique sun-oriented warm transformation property.

Scientists have tried to incorporate various nucleating agents (nanofillers) into the polymer matrix in the present situation in order to advance the dielectric and piezoelectric aspects of this polymer without compromising characteristics like mechanical strength or flexibility [67–69]. Specific interfacial interactions between the polymer chains and nucleating substances affect how polymer functions during crystallisation.

5.5 Application in Energy Renovation and Packing

5.5.1 Energetic solar schemes

The renewal of solar energy into numerous forms of energy, which are then applied, is a key function of energetic solar systems [70]. The energy that is also transformed by these systems is often heat or power. Homes can advantage from active solar systems for heating, cooling and renewable energy generation. The ability of active solar systems to be exploited successfully by a controller could be the most crucial advantage (usually electrical). As among active solar systems that have gained prominence subsequently are dye-sensitised solar cells (DSSCs), solar accumulators, photocatalysts, solar photovoltaics and photocatalysts.

5.5.2 Solar accumulator

It is an effective fluid absorbing heat from a solar accumulator, a form of temperature exchanger that absorbs solar energy. The strength and output temperature [74, 75] of flat-plate solar accumulators, which are strategically located in a wide range of systems, are being enhanced by additional research [71–73]. The performance of the nanofluids, which are nanoparticles (NPs) spread throughout the base fluid [76], enhances daily in solar accumulators due to the increased thermal conductivity of pure fluids. In order to improve the efficiency of solar accumulators and solar water heating devices, nanofluids are useful.

5.5.3 Photocatalysis

Due to their low/less environmental pollution and high efficiency in energy transfer, photocatalysts have experienced extensive development in recent years. The development of TiO_2 is applied in a wide range of crucial areas, which include solar cells, medical symposia, surgical centres for the treatment of cancer and operating in refrigerators or fluorescent lights in electrical applications, containing high levels in water and energy industries. Through the oxidation and reduction of substances to generate pollutants like hydrogen, photocatalytic materials are presently being used to convert solar energy into chemical energy. The pollutants and microorganisms on the wall surfaces also are eliminated by them [77]. Due to their ability for oxidising and degrading organic pollutants, TiO_2 NPs are employed. Due to electrons and holes emerging as a result of their attraction to ultraviolet (UV) light with a bandgap equal to the bandgap, TiO_2 has a photocatalytic property. The activities operate on the exposed TiO_2 particles during UV excitation. DSSCs, lithium-ion sequestrations and electrochromic devices are only a few possibilities for the growing usage of nanostructured TiO_2 materials in photocatalysis [78–81]. TiO_2 nanosheets can serve in self-cleaning glass due to their characteristics, which include a flat surface, moderate turbidity, a high adjacency ratio and specialty area to substrates. They are a potential prospect for a number of applications, such as purification, separation and packaging [82]. Purification, air distillation, water purification, defogging, self-cleaning and hydrogen group are examples of photocatalytic applications.

5.6 Solar and Organic Photovoltaic Cells

The techniques in solar cells are eliminated by transferring solar radiation into more effective electrical energy. Nearly 86% of the global market for solar cells is provided for photovoltaic devices manufactured of inorganic silicon-based materials. Even so, the strength of a solar cell has achieved a phenomenal mark of 20–25%. The photovoltaic technology needed for such a device is too sophisticated, hazardous and demanding, limiting its value and potential to be financially feasible. For many years, research conducted globally has been driven by the findings of various earth-abundant solar cell materials. Conjugated polymers are possible ones to be active in a blend with n-type inorganic semiconductors (typically TiO_2, ZnO and CdSe) to produce informal and cost-effective hybrid organic–inorganic solar cells due to their high coefficient of absorption and durable deposition on functional substrates by simple, cost-effective strategies like lithography and roll-to-roll covering [83–85]. The hybrid polymer of a p-type semiconductor operates simply as a hole transporter and an electron donor. Although so far, the concerts achieved for genuine cells are substantially below these possibilities (1–4%), theoretical recoverable efficiencies of these potential alternatives to silicon-based photovoltaics are anticipated to be as high as 12%. While the combining of organic solar cells has not yet reached the level of commercialisation, optimistic developments over the past few decades have probably increased the prospect to advance substantially from the current state of the art.

5.6.1 Organic photovoltaic solar cell

The photovoltaic effect has the potential to transform photon energy into electricity by generating voltage or electric current in a specific structure when exposed to sunlight [86]. Based on this photovoltaic effect, techniques with various architectures and materials, notably silicon-based solar cells (Savin et al., 2015), thin-film solar cells [87] and third-generation solution-processed solar cells [88], are being explored for solar energy alteration uses. Nowadays, the major of our emphasis is focused toward third-generation solution-processed solar cells, which also include polymer solar cells, dye-sensitised solar cells and perovskite solar cells with polymers functioning as key attributes. In these three devices, the conducting polymer now functions as the electrocatalytic counter electrode, hole-transporting layer and photoactive film.

5.6.2 Dye-sensitised solar cell

In semiconductor photovoltaic systems characterised as dye-sensitised solar cells (DSSC), radiation is initially converted into an electric current. The sensitisation of the inclusive bandgap semiconductors, the configuration of the photoelectrode and the conductor properties, which are pivotal as electrodes

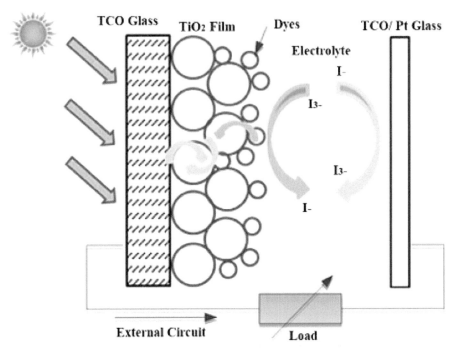

Figure 5.5   Schematic representation of DSSCs.

or hole-conveying materials optically in solution-processed solar cells, all have a major impact on the transformation efficiency of these systems. A DSSC's chemical assembly usually consists of three components a sandwich-like redox electrolyte, a photo anode that is mostly dye-sensitised and a counter electrode. While the counter electrode is essential for redox couple reduction and the dye reduction mechanism, the dye-sensitised photo anode is essential for exposure and charge injection. First, light can traverse through solid electrodes and is attracted to a range of dye in an excited state. The excited electrons would then be transferred into the semiconductor metal oxide's conduction band and transfer to the exterior of the trip. The oxidation variation dye would be reduced by a redox pair in the electrolyte and to complete the circle, the counter electrode would be reduced by exterior circuit electrons. The expensive noble metal counter electrode (Pt) and the corrosive liquid electrolyte are two of the most significant obstacles facing DSSC. Now that conducting polymers have been invented, it is possible to produce better counter electrodes and the use of a poly electrolyte eliminates the necessity for liquid electrolyte, which can precipitate out [89].

5.6.3 Perovskite solar cell

The dye-sensitised solar cell was overtaken by the perovskite solar cell system, which has a superior power conversion efficiency. In less than five years, the efficiency was obtained from 3.8% to a certified 20.2%, illustrating a potential demand in the near future [90]. Due to the inherent properties of this perovskite material, the perovskite solar cell transformed into a simpler structure as compared to its precursor, the dye-sensitised solar cell. The stimulation of a strong conducting electrode, which is constituted of a compact blocking layer (TiO_2 and ZnO), a perovskite light-harvesting layer, a hole-conveying layer and a back interaction electrode, naturally occurs in a high-performance and stable perovskite solar cell. Numerous confessions are obtained as part of the survey. The perovskite light-harvesting material absorbs photons from sunlight with its brightness. Free charge carriers are the amount of potential to excitons because the binding energy between electron–hole pairs is small [91]. In the anode and cathode, respectively, the electrons and holes would be assigned to the layers responsible for carrying them. In order to eliminate and transport holes reliably during a full processing circle, a layer for transporting holes is essential. There will be descriptions of the applications of specific polymer materials as layering techniques for delivering holes.

5.7 Functioning Mechanism

Conjugated semiconducting polymers are used in the development of organic solar systems, which are generated using low-cost solution advancement techniques such as spin-coating, spray-coating and ink-jet printing [92, 93]. The most advanced generation of the functioning OPV-developed induction is the bilayer principle, which comprises a photoactive donor layer and an acceptor layer, in which bilayers (only a few nanometers numerous) are arranged among electrodes. While the anode is comprised of a pure conducting oxide [such as metal tin oxide as indium (ITO)] and interacts with the donor layer, the cathode is made of a relatively low-function metal (such as Al or Ca) and interacts with the acceptor layer. The device's photovoltaic activity is regulated by the bilayer-type dynamic layer and is related to charge storage and sunlight absorption (dropping from the anode side) (Figure 7.1). Donor molecules are semiconductors that, when exposed to light, stimulate photons and produce a small number of electron–hole pairs (excitons) (e.g., CP with a specific bandgap and energy level). The acceptors, on the other hand,

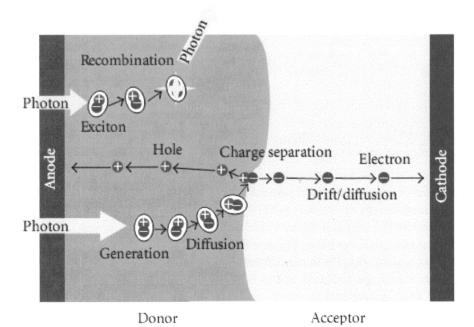

Figure 5.6 Standard of charge parting in a solar cell.

are molecules with a low electron affinity (EA) in reference to the donor, thus they will rapidly take electrons from the excitons present at the donor–acceptor edge (i.e., thorough exciton dissociation). Excitons develop in the donor layer as a result of photon straining in CP.

At the donor–acceptor interface, where exciton separation occurs and electrons are conveyed to the acceptor class, leaving holes in the donor level, the excitons now travel across the thickness of the layer. Then, either directly or via auxiliary electron or hole-transport layers, the electrons and holes are transferred to the relevant electrodes (such as the cathode and anode) to complete the circuitry. The donor conducting polymer is selected based on the active state of the bandgap, hole eminence and energy level, especially the highest occupied molecular orbitals (HOMOs, which have the characteristic of ionisation potential/IP) and smallest vacant molecular orbitals (LUMOs), which is the amount of material EA.

A significant aspect that can result in massive charge generation (due to the collection of more photons) and high current density (due to the development of more free carriers via the separation of parent excitons) in organic photovoltaic devices is broadband absorption, especially in the near-infrared/NIR. The optimum bandgap of polymers for single-junction (bilayer type) organic solar cells is 1.3–1.4 eV [84, 94]. Due to the produced holes' high mobility, the anode can be addressed shortly. To maintain the large cell voltage (or open-circuit voltage) while exciton dissection develops, the HOMO/LUMO energy levels of the donor and acceptor must be precisely aligned with each other, with a sufficient power balance (e.g., 0.3 eV) among them. The donor–acceptor interface, or heterojunction (HJ), is where excitons thrive and are released into free carriers. Excitons are generated within the donor layer and must spread across the thickness of the layer (usually >50 nm for optimal light absorption). The average distance an exciton traverses within the donor layer before ever being trapped, or the exciton diffusion length, is on the order of 10–20 nm for many conducting polymers due to the presence of traps. The consequence is a reduction in carrier absorption and device current due to the fact that excitons produced toward left of the interface (at a distance of >20 nm) are frequently circled and unable to influence the heterojunction. It was proposed that the inherent concerns (such as poor exciton harvesting performance due to exciton trappings) of bilayer (or single heterojunction, Organic photovoltaic devices as heterojunctions) are currently capable throughout the bulk and distributed at a wide scale will be remedied by nanolevel blending (domain size 20 nm) of donor and acceptor to frame co-continuous diffusing sort associations with a significant number of heterojunctions. Bulk heterojunction (BHJ) is a concept that produces a penetrating morphology that used an active layer made up of an intermixed donor and acceptor system.

It is critical to note that for optimal free radical transport, the donor and acceptor phases must be properly mixed to ensure that the donor phase's domain size is between the exciton circulation length of 10–20 nm. P3HT–fullerene derivative (PC60BM, PC70BM and ICBA) nanomaterials (CNT), P3HT–graphene, P3HT–quantum dots and numerous composites made with low-bandgap polymers are instances of donor/acceptor composites.

The development of inorganic semiconductors as photoactive energy harvesters has been assisted by their perfect bandgap, high mobility and moral stability. These hybrid devices are produced by combining the strong optical absorption, variable bandgap/energy layer, simple response degradation and strong hole mobility of conducting polymers with strong electron mobility, strong EA, superior thermal and comprehensive optoelectronic features of NCRs. Solution-processable NCRs provide a substantial zone for robust exciton dissociation when mixing with soluble polymers. In a heterojunction

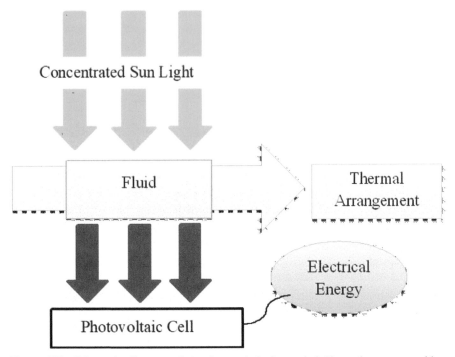

Figure 5.7 Schematic diagram of the de-coupled photovoltaic/thermal system working principles.

device that integrates organic and inorganic mechanisms, conducting polymers act as donors, attracting light and transferring holes, while NCRs act as acceptors, enabling electrons to pass. Naturally, it was found that more inorganic NCRs function as light-harvesting acceptors with functional transport features when blended with conducting polymers. This facilitates them to produce their own transferors, dissociate excitons generated in CPs and successfully transport carriers to the collecting electrode [95]. A power adaptation competence of more than 3% has also been attained in such systems [96].

Recently, inorganic NCRs like ZnO, SnO, TiO_2 and Si were commonly recycled in hybrid solar cells together with CdSe, CdS, PbSe, PbS, CdTe and many others. Researchers aimed to alter NCRs since their size and form have an impact on the bandgap and true energy state. By changing the growth conditions, CdSe can be generated in a variety of forms and sizes, comprising nanoparticles, nanorods and tetrapods [97–99]. TiO_2 nanoparticles, nanorods and CNTs [100, 101], as well as zinc oxide nanorods [104, 105] and nanowires [106–108] can all be produced using chemical synthesis techniques. These NCRs are incorporated into various conducting polymer-based host matrices to obtain their charge separation abilities and organic photovoltaic efficiency [95]. CdSe has now been mixed with several conducting polymers [109] containing MEH-PPV, CN-PPV and MEH-CN-PPV to evaluate the charge transfer from CP to CdSe using photoluminescence (PL) extent. The perfect arrangement of CdSe NCRs has also been evaluated using CP matrices with widely various particle sizes. A basic design criterion for donor polymers in a single-junction hybrid solar cell is to reduce their bandgap to 1.5–1.6 eV with LUMO levels that are Eb higher than the inorganic acceptor CB. The polymer then continues to maintain its V_{OC}, while collecting a spectrum of photons. When choosing the ideal CP/NCR combination for an OPV technology, this could be applied as a design reference to visualise the potential. Inorganic NCRs functioning as electron acceptors in heterojunction solar cell applications must have a stronger EA than donor CPs. It is potentially favorable for exciton dissociation and charge transfer at surfaces whenever CB of acceptor hybrid inorganic semiconductors is lower than LUMO of donor polymers below a diverse range of conditions. Strong electron-accepting ability and electron mobility are also essential for solar cell applications.

It can be noticed that several NCRs have lower conduction bands than the LUMOs of specific CPs, allowing them to act as electron acceptors for those unique CPs and facilitating exciton separation and charge transfer at the CP/NCR heterojunction. It is essential to recognise that while the energy originates from a diversity of donor and acceptor materials, contact dipoles and other variables can lead their levels to modify once they have been combined into surfaces. For this, it is important while designing the device to take into account any mismatch in the energy state driven by boundary errors.

Two main designs for organising hybrid solar cells are the ordered heterojunction (OHJ) and the polymer–inorganic hybrid BHJ [84]. Polymer–inorganic hybrid BHJ solar cells will overcome the difficulties of bilayer devices such constrained donor–acceptor interfacial space and minimal exciton separation, similar to polymer–fullerene BHJ solar cells. To expedite solution processing, NCRs may have their surfaces modified to make them soluble in organic solvents (such as chloroform, solvent and chlorobenzene). The many surface-comforting ligands were applied to change the surface properties of NCRs. Directly combining the two components to produce hybrids or bulk composites or in-situ configuration design one counterpart into the other to generate the required nanocomposites are two different ways to produce the CP/NCR systems (NCs). To develop hybrids with a diffusing interface, TiO_2 would be resolved in the polymer solution during a polymer matrix. To produce organic/inorganic NCs, effective CP monomers with distinct NCRs can also be polymerised in place. In general, OHJ solar cells are recognised as the most capable type for hybrid solar cells due to the direct charge transport routes and well-regulated heterojunctions. Inorganic semiconductors such as nanotubes, quantum dots, nanorods and nanowires can be vertically connected to substrates to produce nanoporous lengths that can be evaluated with CPs to determine the molecular structure of ordered heterojunctions. Due to poor polymer filling driven by CP particles entering the polymer and delaying the nanopores, this technique for producing ordered heterojunction hybrid solar cells is, however, very difficult to achieve. The pore size, relative molecular weight of the polymer, solvent used to combine the polymer and moderating factors all have an effect on the polymer's ability to fill nanopores [110].

New strategies have been proposed to overcome the challenges in considerably filling nanostructured pores with polymers, including in-situ synthesis of the polymers in the pores by UV-light-assisted polymerisation [111], chemical oxidative technique [112] and electro-chemical polymerisation method [113]. Before being introduced to ultra-visible light in an Argon atmosphere, the TiO_2 nanotubes were lordotic in 2,5-diiodothiophene solution. This allowed the C-I bonds produced by the reaction's intermediates to dissociate and the surface-coupled polythiophene to form. The device was developed to evaluate the in situ polymerised polythiophene on TiO_2 nanoparticles to the device concert of polythiophene infiltrated into all those nanotubes. Because of the higher coupling between polythiophene and TiO_2 nanotubes, in-situ polymerised polythiophene above TiO_2 showed a thousand times higher current density than infiltrated polythiophene inside TiO_2, showing effective exciton dissociation.

In-situ NCR manufacturing enables polymers with inorganic nanocrystals to be chemically connected for higher charge transfer from organic polymer to inorganic acceptors. By reacting diethylzinc precursor with P3HT solution, in which it encounters hydrolysis, proceeded by a reduction reaction and hardening the film at 100°C, zinc oxide NRCs had been synthesised within the P3HT polymer matrix [114]. In-situ manufactured ZnO NCRs within the P3HT matrix were used to achieve a device power of 2% at 521 nm with an EQE of 44%.

5.8 Conclusions

The current status of synthetic techniques is the physical, chemical, electrical, optical, and morphological characteristics of polymeric materials, and the development of state-of-the-art energy storage and harvesting systems. Even though there are many possibilities for innovation, now that we are still in the development phase after the rising phase, assets continue to be the major driving assumption. In order to succeed in the energy efficiency stage, the major competition for high-throughput polymer-based manufacturing technologies is fully stated for viability and future research perspectives. There is a great deal of energy all over us. To generate power, the most frequent form of energy used in daily life, nature has its own set of regulations for renovating solar, thermal and vibrational energy. The thermoelectric effect, the only other aspect of the photovoltaic effect that has been explored thus far, still has to be established in order to optimise the structure and functionality of the device.

The CP-based composites provide a great deal in form of electrical property progress, constancy and quality without forfeiting the flexibility and modest technique ability in constructing organic solar cell devices. In this chapter, several PV technologies are examined, as well as the operation of organic solar cells, supercapacitors, solar cell expressions, evaluation parameters, CP composites, polymer–buckminsterfullerenes (C_{60}), polymer–inorganic hybrid and DBA block polymers used in organic photovoltaics. On bilayer and BHJ solar cell architectures, various device structures (conventional and inverted) were indeed allowed. Device coordination and stability concerns and alternatives have also been explored. Out of the various CP-based composites, the polymer/fullerene derivative BHJ composites are the most recognised in the development environment and have achieved maximum cell efficiencies above 10%. Due to inadequate charge transport or reduced composite morphology, other systems using a polymer–hybrid inorganic elements and a DBA block copolymer exhibited high charge transfer, but their device achievement was notably lower than that of the polymer–buckminsterfullerenes composite-based material. To enhance device performance, one of the major competitions is to minimise polymer bandgaps and engineer HOMO/LUMO energy levels. Other chemists have been able to develop novel polymers with reduced bandgap, enhanced HOMO/LUMO energy levels that match the acceptor and substantial hole mobility despite the time and expense for polymer synthesis. The other main challenge is the continuing superiority of single-junction polymer solar in terms of device efficiency. Sequence and multi-junction polymer solar cell topologies using the above diverse substances will be the next area of study focus and direction. It will substantially increase OPV efficiency.

Polymer materials can be developed to be lightweight and flexible and are usually produced at low cost by implementing solution or melting techniques. That implies that flexible and attire electronics, a capable next-generation technology, are particularly well-tailored to polymer materials. For such reasons, more attention should be paid to different ways to progress gadget efficiency by modifying their structural designs. Since polymer nanomaterials are known to be less established than their equivalents in inorganic materials, enhancing the stability of polymer-based devices is extremely important. Energy harvesting devices have so far only been perceived as planar structures and they have been designed thin to enable the flexible nature of modern electronics. Whereas these thin-film devices can be curved, it is challenging to make them asymmetrical in three dimensions.

On the other hand, a snaking deformation is required for a wide range of flexible electrical products. In this regard, a novel family of one-dimensional energy-collecting devices there in shape of fibers is substantial and has recently gained significant benefits. These energy-collecting nanofibers can also be twisted into a diversity of forms without compromising their structural stability. They can also be woven into wearable appliances, a field that is constantly developing and more flexible electronic fabrics. Such fiber-shaped energy harvesting devices have a moderate energy conversion efficiency, making it challenging to provide them on a large scale. The most significant limitation on the rather directions has been observed to be the weak electrical conductivity of the selected fiber electrode. Future research should focus mainly on raising conducting nanofibers and fabrication techniques for fiber devices.

Acknowledgements

We are grateful to the Head, Department of Physics, Rajiv Gandhi University, Itanagar, Arunachal Pradesh for providing facilities for carrying out the work presented here. The authors declare that they have no recognised competing financial interests or particular relationships that could have performed to influence the work reported in this chapter.

References

[1] J. Zhao, A. K. Patwary, A. Qayyum, M. Alharthi, F. Bashir, M. Mohsin, I. Hanif, and Q. Abbas, The determinants of renewable energy sources for the fueling of green and sustainable economy, Energy, 238 (2022) 122029. https://doi.org/10.1016/j.energy.2021.122029

[2] H. Husin, M. Zaki, A critical review of the integration of renewable energy sources with various technologies, Protection and Control of Modern Power Systems, 6(2021)1-18. https://doi.org/10.1186/s41601 -021-00181-3

[3] G. S. Alemán-Nava, V. H. Casiano-Flores, D. L. Cárdenas-Chávez, R. Díaz-Chavez, N. Scarlat, J. Mahlknecht, J. F. Dallemand, R. Parra, Renewable energy research progress in Mexico: A review, Renew. Sustain. Energy Rev. 32 (2014) 140–153. https://doi.org/10.1016/j. rser.2014.01.004

[4] S. Mishra, L. Unnikrishnan, S. K. Nayak, S. Mohanty, Advances in Piezoelectric Polymer Composites for Energy Harvesting Applica- tionsă: A Systematic Review, Macromol. Mater. Eng. 1800463 (2019) 1–25. https://doi.org/10.1002/mame.201800463

[5] G. J. Snyder, E. S. Toberer, Complex thermoelectric materials, Mater. Sustain. Energy a Collect. Peer-Reviewed Res. Rev. Artic. from Nat. Publ. Gr. (2011) 101–110. https://doi.org/10.1142/9789814317665_0 016

[6] H. Peng, X. Sun, W. Weng, X. Fang, Polymer materials for energy and electronic applications. Academic Press; 2016 Sep 1. https://doi.org/ 10.1016/C2015-0-01541-6

[7] S. Patidar, Applications of Thermoelectric Energy: A Review, Int. J. Res. Appl. Sci. Eng. Technol. 6 (2018) 1992–1996. https://doi.org/10 .22214/ijraset.2018.5325.

[8] D. Enescu, Thermoelectric energy harvesting: basic principles and applications. Green energy advances. 2019 Jan 21:1. https://doi.org/ 10.5772/intechopen.83495

[9] G. Sebald, D. Guyomar, A. Agbossou, On thermoelectric and pyro- electric energy harvesting, Smart Mater. Struct. 18 (2009). https: //doi.org/10.1088/0964-1726/18/12/125006.

[10] X. F. Zheng, C. X. Liu, Y. Y. Yan, Q. Wang, A review of thermo- electrics research-Recent developments and potentials for sustainable and renewable energy applications, Renew. Sustain. Energy Rev. 32 (2014) 486–503. https://doi.org/10.1016/j.rser.2013.12.053

[11] H. C. Nayak, S. S. Parmar, R. P. Kumhar, S. Rajput, Modulation in Electric Conduction of PVK and Ferrocene-Doped PVK Thin Films, Electron. Mater., 3 (2022) 53-62. https://doi.org/10.3390/electronicmat3010005

[12] D. Zhao, G. Tan, A review of thermoelectric cooling: materials, modeling and applications, Appl. Therm. Eng. 66 (2014) 15–24. https://doi.org/10.1016/j.applthermaleng.2014.01.074

[13] N. Espinosa, M. Lazard, L. Aixala, H. Scherrer, Modeling a thermoelectric generator applied to diesel automotive heat recovery, J. Electron. Mater. 39 (2010) 1446–1455. https://doi.org/10.1007/s11664-010-1305-2

[14] S. P. Muduli, S. Parida, S. K. Behura, S. Rajput, S. K. Rout, S. Sareen, Synergistic effect of graphene on dielectric and piezoelectric characteristic of PVDF-(BZT-BCT) composite for energy harvesting applications, Polym. Adv. Technol., 33 (2022) 3628-3642. https://doi.org/10.1002/pat.5816

[15] Q. E. Hussain, D. R. Brigham, C. W. Maranville, Thermoelectric exhaust heat recovery for hybrid vehicles, SAE Int. J. Engines. 2 (2009) 1132–1142. https://www.jstor.org/stable/26308459

[16] B. A. Papkin, N. A. Khripach, V. S. Korotkov, D. A. Ivanov, Thermoelectric generator for a vehicle engine cooling system research and development, Int. J. Appl. Eng. Res. 11 (2016) 8557–8564. https://doi.org/10.1016/j.ijft.2021.100063

[17] J. Lofy, L. E. Bell, Thermoelectrics for environmental control in automobiles, In Twenty-First Int. Conf. Thermoelectr. 2002. Proc. ICT'02., 2002: pp. 471–476.urlhttps://doi.org/10.1109/ICT.2002.1190362.

[18] C. A. Howells, Piezoelectric energy harvesting, Energy Convers. Manag. 50 (2009) 1847–1850. https://doi.org/10.1016/j.enconman.2009.02.020.

[19] T. Yucel, P. Cebe, D. L. Kaplan, Structural origins of silk piezoelectricity, Adv. Funct. Mater. 21 (2011) 779–785. https://doi.org/10.1002/adfm.201002077.

[20] E. Fukada, Piezoelectricity of wood, J. Phys. Soc. Japan. 10 (1955) 149–154.https://doi.org/10.1143/JPSJ.10.149

[21] E. Fukada, I. Yasuda, On the piezoelectric effect of bone, J. Phys. Soc. Japan. 12 (1957) 1158–1162. https://doi.org/10.1143/JPSJ.12.1158

[22] D. C. Mountain, A. E. Hubbard, A piezoelectric model of outer hair cell function, J. Acoust. Soc. Am. 95 (1994) 350–354. https://doi.org/10.1121/1.408273

[23] V. Kumar, A. Kumar, S. S. Han, S.-S. Park, RTV silicone rubber composites reinforced with carbon nanotubes, titanium-di-oxide and their hybrid: Mechanical and piezoelectric actuation performance, Nano Mater. Sci. 3 (2021) 233–240.https://doi.org/10.1016/j.nanoms.2 020.12.002

[24] S. S. Rajput, S. Keshri, Structural and microwave properties of (Mg,Zn/Co)TiO_3 dielectric ceramics. J. Mater. Eng. Perform., 23 (2014) 2103-2109.https://doi.org/10.1007/s11665-014-0950-7

[25] B. D. Zaitsev, I. E. Kuznetsova, S. G. Joshi, I. A. Borodina, Shear-horizontal acoustic waves in piezoelectric plates bordered with conductive liquid, IEEE Trans. Ultrason. Ferroelectr. Freq. Control. 48 (2001) 627–631.https://doi.org/10.1109/58.911748

[26] T. Satyanarayana, G. Srinivas, M. S. Prasad, Y. Srinivas, B. Sudheer, K. S. Rao, Design and analysis of MEMS based composite piezoelectric ultrasonic transducer, Electr. Electron. Eng. 2 (2012) 362–373.https: //doi.org/10.5923/j.eee.20120206.04

[27] S. Rajput, X. Ke, X. Hu, M. Fang, D. Hu, F. Ye, X. Ren, Critical triple point as the origin of giant piezoelectricity in $PbMg_{1/3}Nb_{2/3}O_3$-$PbTiO_3$ system. J. Appl. Phys., 128 (2020) 104105. https://doi.org/10 .1063/5.0021765

[28] Sonika, S. K. Verma, S. Samanta, A. K. Srivastava, S. Biswas, R. M. Alsharabi, S. Rajput, Conducting Polymer Nanocomposite for Energy Storage and Energy Harvesting Systems. Adv. Mater. Sci. Eng., 2022 (2022) 1-23. https://doi.org/10.1155/2022/2266899

[29] Sonika, S. K. Verma, J. Saha, R. M. Alsharabi, S. Rajput, Microarticstructural, Spectroscopic, and Magnetic Analysis of Multiwalled Carbon Nanotubes Embedded in Poly (o-aminophenol) Matrices. Adv. Mater. Sci. Eng., 2022 (2022.) 1-9.

[30] L. Mateu, F. Moll, Optimum piezoelectric bending beam structures for energy harvesting using shoe inserts, J. Intell. Mater. Syst. Struct. 16 (2005) 835–845.https://doi.org/10.1177/1045389X0505528

[31] J. Granstrom, J. Feenstra, H. A. Sodano, K. Farinholt, Energy harvesting from a backpack instrumented with piezoelectric shoulder straps, Smart Mater. Struct. 16 (2007) 1810. https://doi.org/10.1088/0964-1 726/16/5/036

[32] B. J. Hansen, Y. Liu, R. Yang, Z. L. Wang, Hybrid nanogenerator for concurrently harvesting biomechanical and biochemical energy, ACS Nano. 4 (2010) 3647–3652.https://doi.org/10.1021/nn100845b

[33] W. Liu, X. Cheng, X. Fu, C. Stefanini, P. Dario, Preliminary study on development of PVDF nanofiber based energy harvesting device for an artery microrobot, Microelectron. Eng. 88 (2011) 2251–2254. https://doi.org/10.1016/j.mee.2011.02.098

[34] D. Koyama, K. Nakamura, Electric power generation using vibration of a polyurea piezoelectric thin film, Appl. Acoust., 71 (2010) 439–445. https://doi.org/10.1016/j.apacoust.2009.11.009

[35] C. M. T. Tien, N. S. Goo, Use of a piezocomposite generating element in energy harvesting, J. Intell. Mater. Syst. Struct. 21 (2010) 1427–1436. https://doi.org/10.1177/1045389X10381658

[36] C. R. Bowen, J. Taylor, E. Leboulbar, D. Zabek, A. Chauhan, R. Vaish, Pyroelectric materials and devices for energy harvesting applications, Energy Environ. Sci. 7 (2014) 3836–3856. https://doi.org/10.1039/c4ee01759e.

[37] A. D. Lantada, P. L. Morgado, H. Hugo, H. Lorenzo-yustos, J. E. Otero, J. M. Munoz-guijosa, J. Muñoz-garcía, J. Luis, M. Sanz, Modelling and Trials of Pyroelectric Sensors for Improving Its Application for Biodevices, In Proceedings of the International Conference on Biomedical Electronics and Devices - Volume 1: Special Session AMMD (BIOSTEC 2009), 453-457, 2009, Porto, Portugal. https://doi.org/10.5220/0001828304530457.

[38] S. S. Rajput, S. Keshri, Structural, vibrational and microwave dielectric properties of $(1-x)$ Mg$_{0.95}$Co$_{0.05}$TiO$_3-(x)$Ca$_{0.8}$Sr$_{0.2}$TiO$_3$ ceramic composites, Journal of alloys and compounds, 581 (2013) 223-229. https://doi.org/10.1016/j.jallcom.2013.05.225

[39] C. Mart, T. Kämpfe, R. Hoffmann, S. Eßlinger, S. Kirbach, K. Kühnel, M. Czernohorsky, L. M. Eng, W. Weinreich, Piezoelectric Response of Polycrystalline Silicon-Doped Hafnium Oxide Thin Films Determined by Rapid Temperature Cycles, Adv. Electron. Mater. 6 (2020) 1–5. https://doi.org/10.1002/aelm.201901015.

[40] H. Zhang, Y. Xie, X. Li, Z. Huang, S. Zhang, Y. Su, B. Wu, L. He, W. Yang, Y. Lin, Flexible pyroelectric generators for scavenging ambient thermal energy and as self-powered thermosensors, Energy, 101 (2016) 202–210. https://doi.org/10.1016/j.energy.2016.02.002.

[41] Y. Yang, H. Zhang, G. Zhu, S. Lee, Z. H. Lin, Z. L. Wang, Flexible hybrid energy cell for simultaneously harvesting thermal, mechanical, and solar energies, ACS Nano, 7 (2013) 785–790. https://doi.org/10.1021/nn305247x.

[42] Y. Li, Molecular Design of Photovoltaic Materials for Polymer Solar Cellsă: Toward Suitable Electronic Energy Levels and Broad Absorption, Accounts Chem. Res., 45 (2012) 723-733. https://doi.org/10.102 1/ar2002446

[43] C. Dun, C. A. Hewitt, H. Huang, J. Xu, D. S. Montgomery, W. Nie, Q. Jiang, D. L. Carroll, Layered Bi_2Se_3 Nanoplate/Polyvinylidene Fluoride Composite Based n-type Thermoelectric Fabrics, ACS Appl. Mater. Interfaces, 7(13), pp.7054-7059.https://doi.org/10.1021/acsami .5b00514

[44] S.Rajput, M. Averbukh, A. Yahalom, T. Minav, An Approval of MPPT Based on PV Cell's Simplified Equivalent Circuit During Fast-Shading Conditions, Electronics, 8 (2019) 1060. https://doi.org/10.3390/electr onics8091060

[45] G. P. Smestad, F. C. Krebs, C. M. Lampert, C. G. Granqvist, K. L. Chopra, X. Mathew, and H. Takakura, Reporting solar cell efficiencies in solar energy materials and solar cells. Sol. Energy Mater. Sol. Cells, 92 (2008) 371-373. https://doi.org/10.1016/j.solmat.2008.01.003

[46] B. C. Brabec, A. Cravino, D. Meissner, N. S. Sariciftci, T. Fromherz, M. T. Rispens, L. Sanchez, J. C. Hummelen, Origin of the open circuit voltage of plastic solar cells. Adv. Funct. Mater., 11 (2001) 374-380. https://doi.org/10.1002/1616-3028(200110)11:5

[47] X. Guo, N. Zhou, S. J. Lou, J. Smith, D. B. Tice, J. W. Hennek, S. Li, J. Strzalka, L. X. Chen, P. Ortiz, J. T. Lo, Polymer solar cells with enhanced fill factors, Nat. Photonics, 7 (2013) 825-833. https://doi.or g/10.1038/nphoton.2013.207.

[48] J. Lv, Y. Cheng, Fluoropolymers in biomedical applications: state-of-the-art and future perspectives, Chemical Society Reviews, 50 (2021) 5435-5467.https://doi.org/10.1039/D0CS00258E

[49] D. Vatansever, R. L. Hadimani, T. Shah, E. Siores, An investigation of energy harvesting from renewable sources with PVDF and PZT, Smart Mater. Struct. 20 (2011). https://doi.org/10.1088/0964-1726/20/5/055 019.

[50] D. Zabek, K. Seunarine, C. Spacie, C. Bowen, Graphene Ink Laminate Structures on Poly(vinylidene difluoride) (PVDF) for Pyroelectric Thermal Energy Harvesting and Waste Heat Recovery, ACS Appl. Mater. Interfaces. 9 (2017) 9161–9167. https://doi.org/10.1021/ac sami.6b16477.

[51] X. Liu, X. Zhao, Y. Yu, Y. Wang, Y. Shi, Q. Cheng, Z. Fang, Y. Yong, Facile fabrication of conductive polyaniline nanoflower modified electrode and its application for microbial energy harvesting, Electrochim. Acta., 255 (2017) 41–47.

[52] T. Badapanda, R. Harichandan, T. B. Kumar, S. Parida, S. S. Rajput, P. Mohapatra, R. Ranjan, Improvement in dielectric and ferroelectric property of dysprosium doped barium bismuth titanate ceramic, J. Mater. Sci.: Mater. Electron., 27 (2016) 7211-7221. https://doi.org/10.1007/s10854-016-4686-z

[53] J. Lu, K. S. Moon, B. K. Kim, C. P. Wong, High dielectric constant polyaniline/epoxy composites via in situ polymerization for embedded capacitor applications, Polymer, 48 (2007) 1510–1516. https://doi.org/10.1016/j.polymer.2007.01.057.

[54] V. Gupta, N. Miura, Polyaniline/single-wall carbon nanotube (PANI/SWCNT) composites for high performance supercapacitors, Electrochim. Acta. 52 (2006) 1721–1726. https://doi.org/10.1016/j.electacta.2006.01.074.

[55] P. Y. Stakhira, O. I. Aksimentyeva, B. R. Cizh, V. V Cherpak, Hybrid solar cells based on dispersed InSe-polyaniline composites, Phys. Chem. Solid State. 6 (2005) 96–98.

[56] Z. Liu, J. Zhou, H. Xue, L. Shen, H. Zang, W. Chen, Polyaniline/ TiO_2 solar cells, Synth. Met. 156 (2006) 721–723. https://doi.org/10.1016/j.synthmet.2006.04.001.

[57] C. Zhao, S. Xing, Y. Yu, W. Zhang, C. Wang, A novel all-plastic diode based upon pure polyaniline material, Microelectronics J. 38 (2007) 316–320. https://doi.org/10.1016/j.mejo.2007.01.004.

[58] K. Ghanbari, M. F. Mousavi, M. Shamsipur, Preparation of polyaniline nanofibers and their use as a cathode of aqueous rechargeable batteries, Electrochim. Acta. 52 (2006) 1514–1522. https://doi.org/10.1016/j.electacta.2006.02.051.

[59] W. Wang, E. A. Schiff, W. Wang, E. A. Schiff, Polyaniline on crystalline silicon heterojunction solar cells Polyaniline on crystalline silicon heterojunction solar cells, Appl. Phys. Lett., 91 (2007)133504. https://doi.org/10.1063/1.2789785.

[60] M. Khalifa, S. Anandhan, PVDF Nanofibers with Embedded Polyaniline – Graphitic Carbon Nitride Nanosheet Composites for Piezoelectric Energy Conversion, ACS Appl. Nano Mater., 2 (2019) 7328-7339. https://doi.org/10.1021/acsanm.9b01812.

[61] U. Johanson, A. Punning, A. Aabloo, Ionic polymer metal composites with electrochemically active electrodes, RSC Smart Mater. 2016 (2016) 215–227. https://doi.org/10.1039/9781782622581-00215.

[62] V. F. Ferreira, M. C. B. V de Souza, A. C. Cunha, L. O. R. Pereira, M. L. G. Ferreira, Recent advances in the synthesis of pyrroles, Org. Prep. Proced. Int. 33 (2001) 411–454.

[63] M. Bharti, A. Singh, S. Samanta, A. K. Debnath, D. K. Aswal, K. P. Muthe, S. C. Gadkari, Flexo-green Polypyrrole-Silver nanocomposite films for thermoelectric power generation, Energy Convers. Manag. 144 (2017) 143–152.

[64] M. Culebras, B. Uriol, C. M. Gómez, A. Cantarero, Controlling the thermoelectric properties of polymers: Application to PEDOT and polypyrrole, Phys. Chem. Chem. Phys. 17 (2015) 15140–15145. https://doi.org/10.1039/c5cp01940k.

[65] J. Xu, Y. Tan, X. Du, Z. Du, X. Cheng, H. Wang, Cellulose nanofibril/polypyrrole hybrid aerogel supported form-stable phase change composites with superior energy storage density and improved photothermal conversion efficiency, Cellulose. 27 (2020) 9547–9558. https://doi.org/10.1007/s10570-020-03437-7

[66] M. Silakhori, H. Fauzi, M. R. Mahmoudian, H. S. C. Metselaar, T. M. I. Mahlia, H. M. Khanlou, Preparation and thermal properties of form-stable phase change materials composed of palmitic acid/polypyrrole/graphene nanoplatelets, Energy Build., 99 (2015) 189–195. https://doi.org/10.1016/j.enbuild.2015.04.042

[67] A. Singh, A. Sharma, S. Rajput, A. K. Mondal, A. Bose, and M. Ram, Parameter Extraction of Solar Module Using the Sooty Tern Optimization Algorithm, Electronics, 11 (2022) 564. https://doi.org/10.3390/electronics11040564

[68] B. Ponraj, R. Bhimireddi, K. B. R. Varma, Effect of nano-and micron-sized $K_{0.5}Na_{0.5}NbO_3$ fillers on the dielectric and piezoelectric properties of PVDF composites, J. Adv. Ceram., 5 (2016) 308-320. https://doi.org/10.1007/s40145-016-0204-2

[69] A. Singh, A. Sharma, S. Rajput, A. Bose, and X. Hu, An Investigation on Hybrid Particle Swarm Optimization Algorithms for Parameter Optimization of PV Cells, Electronics, 11 (2022)909. https://doi.org/10.3390/electronics11060909

[70] A. Mohammadi, M. Hossein, M. Bidi, M. Ghazvini, Exergy and economic analyses of replacing feedwater heaters in a Rankine cycle

with parabolic trough collectors, Energy Reports, 4 (2018) 243–251. https://doi.org/10.1016/j.egyr.2018.03.001.

[71] D. Rojas, J. Beermann, S. A. Klein, D. T. Reindl, Thermal performance testing of flat-plate collectors, Solar Energy, 82 (2008) 746–757. https://doi.org/10.1016/j.solener.2008.02.001.

[72] S. Rajput, A. Lugovskoy, M. Averbukh, A. Yahalom, Porous Metal-Oxide Based Electrostatic Energy Generator. In 2019 International IEEE Conference and Workshop in Óbuda on Electrical and Power Engineering (CANDO-EPE) (pp. 133-136). https://doi.org/10.1109/CANDO-EPE47959.2019.9110961

[73] R. Tang, Y. Yang, W. Gao, Comparative studies on thermal performance of water-in-glass evacuated tube solar water heaters with different collector tilt-angles, Sol. Energy. 85 (2011) 1381–1389. https://doi.org/10.1016/j.solener.2011.03.019.

[74] K. Y. Leong, H. Chyuan, N. H. Amer, M. J. Norazrina, M. S. Risby, K. Z. K. Ahmad, An overview on current application of nano fl uids in solar thermal collector and its challenges, Renew. Sustain. Energy Rev. 53 (2016) 1092–1105. https://doi.org/10.1016/j.rser.2015.09.060.

[75] F. S. Javadi, R. Saidur, M. Kamalisarvestani, Investigating performance improvement of solar collectors by using nano fl uids, Renew. Sustain. Energy Rev. 28 (2013) 232–245. https://doi.org/10.1016/j.rser.2013.06.053.

[76] A. Sharma, A. Sharma, V. Jately, M. Averbukh, S. Rajput, and B. Azzopardi, A Novel TSA-PSO Based Hybrid Algorithm for GMPP Tracking under Partial Shading Conditions, Energies, 15 (2022) p. 3164. https://doi.org/10.3390/en15093164

[77] L. Zhang, Y. Ding, M. Povey, D. York, ZnO nanofluids – A potential antibacterial agent, Prog. Nat. Sci., 18 (2008) 939-944. https://doi.org/10.1016/j.pnsc.2008.01.026.

[78] K. Nakata, A. Fujishima, TiO_2 photocatalysis: Design and applications, J. Photochem. Photobiol. C Photochem. Rev. 13 (2012) 169–189.

[79] S. Rajput, M. Averbukh, N. Rodriguez, Energy Harvesting and Energy Storage Systems. Electronics, 11 (2022) 984.

[80] D. V Bavykin, J. M. Friedrich, F. C. Walsh, Protonated titanates and TiO_2 nanostructured materials: synthesis, properties, and applications, Adv. Mater. 18 (2006) 2807-2824.

[81] S.-J. Bao, C. M. Li, J.-F. Zang, X.-Q. Cui, Y. Qiao, J. Guo, New nanostructured TiO_2 for direct electrochemistry and glucose sensor applications, Adv. Funct. Mater., 18 (2008) 591–599.

[82] K. Nakata, A. Fujishima, Photochemistry Reviews TiO$_2$ photocatalysis: Design and applications, J. Photochem. Photobiol. C, 13 (2012) 169–189. https://doi.org/10.1016/j.jphotochemrev.2012.06.001.

[83] W. U. Huynh, J. J. Dittmer, A. P. Alivisatos, Hybrid nanorod-polymer solar cells, Science, 295 (2002) 2425–2427.

[84] X. Hu, S. Rajput, S. Parida, J. Li, W. Wang, L. Zhao, X. Ren, Electrostrain Enhancement at Tricritical Point for BaTi$_{1-x}$Hf$_x$O$_3$ Ceramics. J. Mater. Eng. Perform., 29 (2020) 5388-5394.https://doi.org/10.1007/s1 1665-020-05003-5

[85] L. Zhao, Z. Lin, Crafting semiconductor organic-inorganic nanocomposites via placing conjugated polymers in intimate contact with nanocrystals for hybrid solar cells, Adv. Mater., 24 (2012) 4353–4368.

[86] C. W. Tang, Two-layer organic photovoltaic cell, Appl.Phys.Lett., 48 (1986) 183-185. https://doi.org/10.1063/1.96937.

[87] J. D. Major, R. E. Treharne, L. J. Phillips, K. Durose, A low-cost non-toxic post-growth activation step for CdTe solar cells, Nature, 511 (2014) 334–337. https://doi.org/10.1038/nature13435.

[88] J. Yan, B. R. Saunders, RSC Advances Third-generation solar cellsă: a review and comparison of polymer: fullerene, hybrid polymer and perovskite solar cells, RSC Adv., 4 (2014) 43286–43314. https://doi.or g/10.1039/C4RA07064J.

[89] K. Saranya, A. Subramania, Developments in conducting polymer based counter electrodes for dye-sensitized solar cells - An overview, Eur. Polym. J.,66 (2015) 207-227. https://doi.org/10.1016/j.eurpolym j.2015.01.049.

[90] A. Sharma, A. Sharma, M. Averbukh, S. Rajput, V. Jately, S. Choudhury, B. Azzopardi, Improved moth flame optimization algorithm based on opposition-based learning and Lévy flight distribution for parameter estimation of solar module, Energy Reports, 8 (2022) 6576-6592. https://doi.org/10.1016/j.egyr.2022.05.011

[91] S. Collavini, S. F. Vçlker, J. L. Delgado, Understanding the Outstanding Power Conversion Efficiency of Perovskite-Based Solar Cells, Angew. Chem. Int. Ed., 54 (2015) 9757–9759. https://doi.org/10.103 8/srep00591.

[92] G. C. Bakos, Feasibility study of a hybrid wind/hydro power-system for low-cost electricity production, Applied Energy, 72 (2002) 599–608. https://doi.org/10.1016/S0306-2619(2)00045-4.

[93] S. Rajput, M. Averbukh, A. Yahalom, Electric power generation using a parallel-plate capacitor, Int. J. Energy Res, 43 (2019) 3905-3913. https: //doi.org/10.1002/er.4492

[94] K. K. Sahoo, S. S. Rajput, R. Gupta, A. Roy, A. Garg, Nd and Ru co-doped bismuth titanate polycrystalline thin films with improved ferroelectric properties, J. Phys. D: Appl. Phys., 51 (2018) 055301. https://doi.org/10.1088/1361-6463/aa9fa5

[95] M. M. Wienk, M. G. R. Turbiez, M. P. Struijk, M. Fonrodona, R. A. J. Janssen, M. M. Wienk, M. G. R. Turbiez, M. P. Struijk, Low-band gap poly (di-2-thienylthienopyrazine): fullerene solar cells, Appl. Phys. Lett., 88 (2006) 153511. https://doi.org/10.1063/1.2195897.

[96] S. Dayal, N. Kopidakis, D. C. Olson, D. S. Ginley, G. Rumbles, Photovoltaic Devices with a Low Band Gap Polymer and CdSe Nanostructures Exceeding 3 % Efficiency, Nano letters, 10 (2010) 239–242. https://doi.org/10.1021/nl903406s.

[97] Z. A. Peng, X. Peng, Formation of High-Quality CdTe, CdSe, and CdS Nanocrystals Using CdO as Precursor, J. Am. Chem. Soc., 123 (2001), 183-184. https://doi.org/10.1021/ja003633m

[98] S. Keshri, S. Rajput, S. Biswas, L. Joshi, W. Suski, P. Wiśniewski, Structural, magnetic and transport properties of Ca and Sr doped Lanthanum manganites,J. Met. Mater. Miner., 31 (2021) 62-68.https://doi.org/10.14456/jmmm.2021.58

[99] A. Ghani, S. Yang, S. Rajput, S. Ahmed, A. Murtaza, C. Zhou, X. Song, Tuning the conductivity and magnetism of silicon coated multiferroic $GaFeO_3$ nanoparticles. J. Sol-Gel Sci. Technol., 92 (2019) 224-230.https://doi.org/10.1007/s10971-019-05096-y

[100] P. D. Cozzoli, A. Kornowski, H. Weller, Low-Temperature Synthesis of Soluble and Processable Organic-Capped Anatase TiO_2 Nanorods, Journal of the American chemical society, 125 (2003) 14539–14548. https://doi.org/10.1021/ja036505h

[101] X. Wang, J. Zhuang, Q. Peng, Y. Li, A general strategy for nanocrystal synthesis, Nature, 437 (2005) 121–124. https://doi.org/10.1038/nature03968.

[102] S. M. Liu, L. M. Gan, L. H. Liu, W. D. Zhang, H. C. Zeng, Synthesis of Single-Crystalline TiO_2 Nanotubes, Chemistry of materials, 14 (2005) 1391–1397.https://doi.org/10.1021/cm0115057

[103] J. H. Park, S. Kim, A. J. Bard, Novel Carbon-Doped TiO 2 Nanotube Arrays with High Aspect Ratios for Efficient Solar Water Splitting, Nano letters, 6 (2006) 24-28. https://doi.org/10.1021/nl051807y

[104] Y. Xie, P. Joshi, S. B. Darling, Q. Chen, T. Zhang, D. Galipeau, Q. Qiao, Electrolyte Effects on Electron Transport and Recombination

at ZnO Nanorods for Dye-Sensitized Solar Cells, The Journal of Physical Chemistry C, 114 (2010) 17880–17888.https://doi.org/10.1021/jp106302m

[105] C. Hung, W. Whang, A novel low-temperature growth and characterization of single crystal ZnO nanorods, Materials Chemistry and Physics 82 (2003) 705–710. https://doi.org/10.1016/S0254-0584(3)00331-6.

[106] A. M. Peiro, P. Ravirajan, K. Govender, D. S. Boyle, P. O. Brien, D. D. C. Bradley, J. R. Durrant, Hybrid polymer / metal oxide solar cells based on ZnO columnar structures,Journal of Materials Chemistry, 16 (2006) 2088-2096. https://doi.org/10.1039/b602084d.

[107] S. Zhang, Y. Shen, H. Fang, S. Xu, Z. Lin, Growth and replication of ordered ZnO nanowire arrays on general flexible substrates, Journal of Materials Chemistry 20 (2010) 10606–10610. https://doi.org/10.1039/c0jm02915g.

[108] L. E. Greene, B. D. Yuhas, M. Law, D. Zitoun, P. Yang, Solution-Grown Zinc Oxide Nanowires, Inorganic chemistry, 45 (2006) 4977–4984. https://doi.org/10.1021/ic0601900

[109] D. S. Ginger, N. C. Greenham, Charge separation in conjugated-polymer/nanocrystal blends, Synthetic Metals, 101 (1999) 425-428. https://doi.org/10.1016/S0379-6779\(98)00330-0

[110] J. Boucle, J. Nelson, Hybrid polymer – metal oxide thin films for photovoltaic applications, Journal of Materials Chemistry 17 (2007) 3141–3153. https://doi.org/10.1039/b706547g.

[111] S. Tepavcevic, S. B. Darling, N. M. Dimitrijevic, T. Rajh, J. Sibener, Improved Hybrid Solar Cells via in situ UV Polymerization, Small, 5 (2009) 1776–1783. https://doi.org/10.1002/smll.200900093.

[112] S. Dwivedi, H. C. Nayak, S. S. Parmar, R. P. Kumhar, S. Rajput, Calcination temperature reflected structural, optical and magnetic properties of nickel oxide. Magnetism, 2 (2022) 45-55. https://doi.org/10.3390/magnetism2010004

[113] Y. Hao, J. Pei, Y. Wei, Y. Cao, S. Jiao, F. Zhu, J. Li, D. Xu, Efficient Semiconductor-Sensitized Solar Cells Based on Poly (3-hexylthiophene)@CdSe@ZnO Core-Shell Nanorod Arrays, J. Phys. Chem. C, 114 (2010) 8622–8625. https://doi.org/10.1021/jp911263d

[114] S. D. Oosterhout, M. M. Wienk, S. S. Van Bavel, R. Thiedmann, L. J. A. Koster, J. Gilot, J. Loos, V. Schmidt, R. A.J. Janssen, the efficiency

of hybrid polymer solar cells, Nat. Mater. 8 (2009) 818–824. https: //doi.org/10.1038/nmat2533.

[115] H. Savin, P. Repo, G. Gastrow, P. Ortega, E. Calle, M. Garin, R. Alcubilla, Black silicon solar cells with inter-digitated back contacts achieve 22.1% efficiency, Nat. Nanotech. 10 (2015) 624–628.

6

Fabrication of Piezoelectric Material and Energy Harvesting While Walking

Sakshi Chauhan and Neeraj Bisht

Department of Mechanical Engineering, College of Technology,
GBPUAT, India
E-mail: sakshichauhan9may@gmail.com; neerajbisht30@gmail.com

Abstract

In the present era of environmental concerns, it is required to look for greener resources for energy harvesting. Many government agencies are projecting their vision of a high proportion of renewable and cleaner energy as the main energy suppliers. Piezomaterials are one such resource that helps to generate electricity. The principle of piezoelectric materials works by applying mechanical stress to a crystal, which can generate a voltage or potential energy difference and thus a current. The present chapter aims to design the experimental set-up to harvest energy using piezo material such as Rochelle salt by primarily synthesising using sodium carbonate and potassium bi-tartrate. The same material is then fabricated into a shoe sole to generate sufficient energy to charge mobile phones portably. In order to generate a sufficient amount of voltage by the application of stress the number of piezo material plates are used.

Keywords: Piezoelectric Materials; Energy Harvesting; Clean Energy; Shoe Sole.

6.1 Introduction

There is currently an enormous energy concern. The International Energy Agency's New Policies Scenario estimates global power demand to increase by about 80% for the period 2012–2040. Additionally, the percentage of renewable energy in total power generation will increase from 21% in 2012 to 33% by 2040 (source: International Energy Agency). The rapidly growing proportion and relevance of renewable energy in the power production sector have resulted in the development of a number of sustainable technologies [1].

As we look for greener and more efficient solutions, the need for energy-harvesting technology is increasing. Piezoelectricity, like wind turbines and solar cells, is a form of energy-collecting technology. Piezoelectricity is the production of electrical energy by the application of mechanical pressure through walking, etc. [2, 3]. When you apply pressure to a material, a negative charge is developed on the expanded side, while the opposite charge is generated on the compressed side. An electric current runs across the material when the pressure is alleviated, as shown in Figure 6.1. Piezoelectric energy channelising has emerged as the technique of choice for powering meso-to-micro size devices out of a variety of viable energy harvesting methods [4]. Piezoelectricity is an area that is currently undeveloped but has a lot of promise in the future. It is widely used in strain gauges, acoustic emission detection and other applications. Piezoelectricity is the charge that develops in solid materials, most notably crystals and some ceramics, as well as biological materials, such as bone, DNA and other proteins, when mechanical strain is applied. Energy harvesting may be accomplished using piezoelectric materials and transducers that can take a broad variety of input frequencies and pressures [5]. Because nano-triboelectric generator materials are not often accessible on the market, our research focuses on piezoelectric energy harvesters.

The potassium sodium tartrate tetrahydrate, commonly called Rochelle salt, is a double salt of tartaric acid that was invented (about 1675) by a French pharmacist named Pierre Seignette. It was the first piezoelectric substance to be identified. This property led to its widespread usage in post-World War II 'crystal' gramophone pick-ups, microphones and earpieces [6].

The power output of a piezoelectric energy harvester is dependent on extrinsic and intrinsic parameters. Intrinsic variables include the piezoelectric element's frequency constant, material characteristics, temperature and stress dependency on physical parameters. The input vibration frequency, the acceleration of the host structure and the magnitude of the stimulation

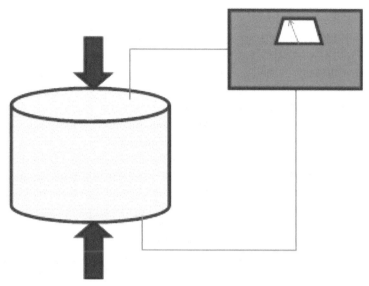

Figure 6.1 Schematic diagram showing the generation of voltage via piezoelectric on deformation.

are all extrinsic parameters. Most of these devices employ either a layer of a piezoelectric material coupled to a non-piezoelectric layer (unimorph) or a bilayer of a piezoelectric material coupled to a non-piezoelectric layer (bimorph) [7].

Due to their better performance, lead zirconate titanate (PZT) and polyvinylidene difluoride (PVDF) are the most commonly used materials for energy harvesting. PZT is hard, brittle and heavy, which limits its use in wearable applications that require flexibility, while PVDF has a high degree of flexibility and strong stability and is simple to handle and mould. PVDF is better suited for wearable applications than PZT because of its high-amplitude and low-frequency human motion attributes [8].

The goal of this research is to synthesise Rochelle salt from sodium carbonate and potassium bi-tartrate and then analyse the salt using a cathode ray oscilloscope (CRO). Furthermore, to design a setup that contains a few piezoelectric sensors that will be able to generate a sufficient amount of voltage by the application of stress on them. The project will be designed such that its application will be user-friendly, flexible and can be used in many applications accordingly. Also, the output current generated will be of sufficient magnitude in order to work efficiently.

6.2 Review of Literature

In this section, a literature survey on the various aspects has been investigated to quantify the amount of work done and its direction in the domain of piezoelectric materials. All the required and nearby areas of research are critically studied in order to find the correct gap in the literature available. This section aims to provide an understanding of piezoelectricity, devices developed and issues raised. In addition, various testing methodologies have been discussed in order to fulfil the required objectives.

A substantial amount of research has been conducted in the field of clean and reliable resources. The use of piezoelectric materials has been mentioned in a limited amount of literature in the field of portable devices. PVDF has been used to develop wearable devices such as shoes, backpacks and apparel. Kymissis et al. created an insole composed of eight layers of 28-micron PVDF sheets which had a 2 mm flexible plastic substrate in the centre. An average output of 1.1 mW at 1 Hz was obtained [3].

PVDF straps were developed as backpack shoulder straps, it was observed that an average power of 45.6 mW during a walking speed of 0.9–1.3 m/s was obtained [9]. A PVDF shell structure trying to capture energy from human motion has also been developed. An output power of 0.87 mW was obtained [10].

Earlier advances in piezoelectric technology focused on micro-vibrations, which produced minor strains. Also, the external voltage had to be supplied. There were a lot of losses in the system, as a result of which a very low voltage output was obtained. The United States Navy has also investigated the utility of piezoelectric crystals. The primary focus of the study was on crystal dimensions [11]. Their study demonstrated that altering the diameter and orientation of the crystal has a significant effect on the output. 'Curie Cut' or 'Zero Cut' crystals were created. This means that crystals of this size can effectively control the oscillations of a 50-watt vacuum tube and can also be used to control voltage.

In 2000, several uses of piezoelectricity in wireless detection were extensively evaluated. Various military and industrial applications necessitate remote sensing of multiple pieces of equipment and machinery. Equipment operating parameters in regions where conventional power sources are inaccessible and prolonged periods of uninterrupted functioning are required are challenges which can be checked using such techniques. Nevertheless, some source of vibrating energy is always present in the operation of the machine, which can be further utilised. They can be efficiently used to generate power for the workings of the microcontroller and transmitters. Several strategies/technologies for the effective conversion, usage and storage of piezoelectric power are identified and utilised in the development of a generic energy-harvesting data transmitter. Based on this approach, the channelised vibrational energy of turbines was harnessed to develop piezoelectric nanogenerators (VT-PENG). These allow energy harvesting in amphibious conditions. They also worked on the geometry of the piezoelectric strips used. The maximum power harvested varies from 0.4 to 0.9 uW for curved piezoelectric strips and planar strips, respectively [12].

In 2005, the United States Defence Advanced Research Projects Agency (DARPA) launched an innovative study on energy harvesting that aims to power combat equipment using piezoelectric generators installed in soldiers' boots [13]. These energy-collecting sources, however, have an effect on the human body. DARPA's attempts to get 1–2 watts from shoes that always touched the ground while walking was thrown out because the extra energy used by the person wearing the shoes made them feel uncomfortable.

Piezoelectric ceramics, which are ferroelectric materials, have been studied extensively. They are polar crystals; hence an electric field is not required for their work. The V–I characteristics of PZT and PVDF were plotted to investigate the output when different types of forces were applied. The piezo transducer component under investigation is positioned on a piezo force sensor for this purpose. Voltmeters are linked across them to measure voltages, and an ammeter is attached to monitor the current. Voltage and current were measured for each voltage reading [3].

Significant technical developments in miniaturised sensors and systems have to be made applicable to embedded wearable technologies. The power supplies' comparatively large size, weight and constrained lifespan are a barrier to the broad deployment of these micro-systems. Emerging micro-power sources use generators that run on their own power and take advantage of the way smart materials naturally convert energy [14].

Scientists have begun to investigate techniques for extracting energy from ambient sources in order to extend the lifecycle of instruments. Advances in this domain have resulted in the creation of various methods that may be utilised to make electrical energy from various mechanical phenomena. Many of these energy sources can be used by individuals, but their use must be taken into account carefully in order to avoid parasite effects that might alter the user's gait or tolerance [15]. Previous experiments to embed these technologies into wearable devices so that the energy emitted during hits may be gathered have resulted in various challenges.

Recently, several researchers worked on piezoelectric generators and reported their sensitivity to minute strain, and there was an increase of three to five times in energy production. In addition, they were more durable and reliable in comparison to conventional methods of energy harvesting [16]. Moreover, the miniaturisation or compact structures of these generators can be easily integrated into micro-electro-mechanical systems [17].

Phononic crystals [18, 19], electrospun fibres and inorganic nanowires are also emerging as possible alternatives for energy harvesting [20]. In 2022, one of the interesting papers reported energy harvesting from fluid-conveying piezoelectric pipes. These pipes have self-powered sensing. It also discusses the various aspects affecting energy generation, like fluid velocity [21].

'Energy harvesting' is the process of extracting, converting and storing energy from nature. Through their intrinsic piezoelectric effect, they have been employed to transduce mechanical energy to electrical energy. The present work focuses on the production of Rochelle salt as well as the fabrication of a piezoelectric plate-based walking charger. The purpose of this research is to create a new type of energy-harvesting shoe that will produce electrical energy. The purpose of this technology is to harness the energy from the shoe and at the same time not affect his or her performance. A new approach could be the replacement of the conventional shoe insole with a piezoelectric polymer-based lead zirconate titanate (PZT) insole [22]. Piezoelectric materials have a configuration that causes mechanical strain when an electrical potential is applied. In contrast, applied stress creates an electrical charge, making it suitable for energy harvesting purposes. PZT is flexible as well as strong, allowing it to successfully serve as the load-carrying part. To retain the efficacy of the shoes and the user, the shoe design will be kept as close to existing systems as feasible.

6.3 Materials and Methods

6.3.1 Printed circuit board (PCB)

Printed circuit board fabrication is the primary stage in any electronic industry's manufacturing industry. The easiest way of manufacturing a PCB is to create a design on a copper board with ink, and then conduct the etching procedure in which the rest of the copper is dissolved in acid. The dirt is scrubbed from the copper sheet using spirit to eliminate traces of grease or oil, after which the board is washed using water. The surface is then dried by a forced jet of heated air or by letting it dry naturally for some period of time.

The development of a PCB drawing requires certain important considerations, like the thickness of lines and holes based on the components. Furthermore, etching the liquid onto the circuit schematic according to the drawing of the PCB (design tracks, rows and squares). Drill the designated holes in the PCB using 1 mm drill bits after the surface gets completely dry. If there are any short lines as a result of the paint pouring, they can be erased once the paint has dried by peeling them off using any pointed device.

After drying, boil 22–30 grams of ferric chloride in 75 ml of water at roughly 60 °C is done. This is then poured over the PCB, on the copper side up, in a 15×20 cm plastic tray. Stirring the solution speeds up the etching process. Unwanted copper would dissolve in roughly 45 minutes. If the etching process takes too long, the solution can be heated again and the procedure may be repeated. The pattern's paint was removed, and the PCB was washed and dried. To keep the gloss, a layer of varnish was applied.

6.3.2 Diode

The sandwich of p-type semiconducting material with contacts connecting the p and n-type layers to an external circuit, which is known as a junction diode, is most commonly used. High current flows if the positive terminal of the battery is linked to the p-type material (cathode) and the negative terminal is connected to the n-type material (anode), resulting in forward bias. The amount of current flow reduces significantly if the connections are reversed. This can be attributed to the fact that, under these conditions, the p-type material absorbs electrons from the battery's negative terminal while the n-type material gives up its free electrons to the battery, and a state of electric equilibrium is reached. As a result, a very meagre current will flow, and the diode is said to be reverse biased. Power diodes convert alternating current (AC) to direct current (DC). In this case, current easily flows during the first half cycle (forward bias), zero current flows during the second half cycle (reverse bias) and the diode functions as an effective rectifier. In the circuit, Zener diodes regulate the voltage. When a junction diode is forward biased, energy is released; this energy is then released by silicon and germanium diodes.

6.3.3 Soldering kit

The soldering equipment was utilised to connect all of the terminals in the circuit. It consists of the various elements. Soldering is the process of putting two metallic pieces together, and the tool utilised for this is known as a soldering iron. Its purpose is to melt solder and assemble metal pieces together. Soldering irons are classified by their wattage, ranging from 10 to 200 watts. Solder is made up of lead and tin. A high-quality solder (a type of flexible bare wire) is 60% tin and 40% lead and melts at temperatures ranging from 180 to 200 °C. The oxide coating that forms when solder spots are formed must be removed immediately. This is achieved by applying a flux, which boils under the heat and eliminates oxide, thereby allowing metal and solder to join. This typical tool is used to clean the surface and terminals of solderable components. To avoid the oxide development attacking the tip, cleaning of the tip with sand paper on a regular basis must be done. Apart from these tools, the soldering workbench also contains other accessories.

6.3.4 Salt

Sodium carbonate, also known as washing soda, is a water-soluble salt of carbonic acid. Sodium carbonate, in its purest form, is a hygroscopic powder that is white in colour and has no odour. It has a highly alkaline flavour and dissolves in water to generate a mildly basic solution. Additionally, potassium bitartrate, commonly known as potassium hydrogen tartrate or cream of tartar, is a tartaric acid potassium salt. It may be used in baking as well as cleaning.

6.3.5 Piezoelectric crystals

Rochelle salt crystals created generate little voltage. Therefore, piezoelectric ceramic plates are required to transduce the applied mechanical stress into electrical energy. A piezo disc with an outside diameter of 2.7 cm and an inner diameter of 2 cm is employed.

6.4 Methodology

The aim of this chapter is to present the experimental techniques used in the fulfilment of the objectives for the present chapter. Various techniques are used to develop materials, devices and access properties of devices under consideration. Therefore, the following sections in detail explain the synthesis of piezoelectric material and its assembly in the shoe sole in order to develop a portal walking charger.

6.4.1 Synthesis of Rochelle salt

Rochelle salt crystals were synthesised using 500 grams of washing soda (sodium carbonate, Na_2CO_2), 200 grams of cream of tartar (potassium bitartrate, $KHC_4H_2O_6$) and 250 ml of distilled water. In 250 ml of water, dissolve 200 g of cream of tartar, and continuously stir the water to ensure that the cream of tartar particles is suspended. The container in a larger pot than the one used in the previous step is placed on the stove and filled halfway with water. The water in the saucepan was heated until it was about to simmer. Half a teaspoon, i.e., approximately 2.5 ml of sodium carbonate, was added to the container, which will float to the top consequently. Keep stirring it until it is no longer bubbling or has returned to its original level (Figure 6.1 (a)). Add the additional sodium carbonate and mix once more. Using the coffee filter, transfer the hot solution from the first container to another container. Once everything is in the second container, continue to heat the filtered solution until less than what was started is left (Figure 6.1 (b)). The purpose of this is to concentrate the solution, after which the solution is kept in a cool area for several days [23].

Since the Rochelle salt is manufactured by hand, it cannot produce more than 10 mV. Therefore, it is required to use the piezoelectric ceramic plates in the walking charger. Voltage by Rochelle salt is produced to a limit of 10 mV while piezoelectric ceramics can produce up to 2 V, found through set up as seen in Figure 6.2. Piezoelectric ceramic plates were developed by combining fine powders of metal oxide in predetermined amounts and heating them. The powder, thereafter, is mixed with an organic binder and turned into the desired shapes. They are then burned according to a precise schedule

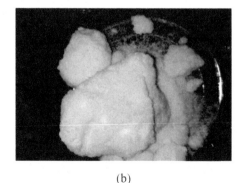

(a) (b)

Figure 6.1(a) (a) Process of Rochelle salt synthesis and (b) synthesised Rochelle salt.

Figure 6.2   The voltage generated for a single impact on 1 piezo crystal.

of time and temperature, resulting in the sintering of powder particles into sinter and the material. After cooling, the elements are machined according to specifications. Electrodes and elements are linked to each other by a conducting substance.

6.4.2 Making of walking charger

6.4.2.1 Principle

The fundamental concept at work is the conversion of human mechanical power into electrical signals. When a person walks or runs, he dissipates kinetic energy and so exerts some pressure on the ground via the shoe; this is the pressure involved and is employed in product development. This vibration induced mechanical energy is fed as an input to the piezoelectric disc transducer, causing the piezo crystals to vibrate, creating an excitation voltage measured in reference to the base plate. The golden layer is the ground plate in Figure 6.3 (a), and the crystals are implanted in the middle as illustrated in Figure 6.3 (b) [8]. The voltage yield of this transducer is then attached to a rectifier, and it is then synchronised to a value adequate enough to charge a mobile phone battery. The voltage this transducer gives off is then hooked up to a rectifier and set to a level that can charge a cell phone battery.

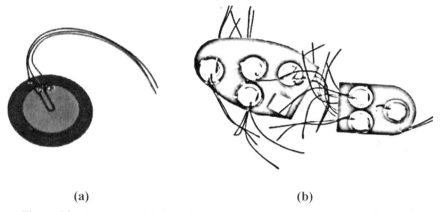

(a)   (b)

Figure 6.3 (a) Piezoelectric transducer disc and (b) connections made in the insole.

6.4.2.2 Design and development

A piezoelectric crystal's output is an alternating signal; therefore, voltage must first be transformed into a DC signal before it can be used for charging. To convert AC to DC, a basic diode rectifier is utilised, which is followed by a capacitor, which enhances the voltage profile generated. The resulting direct voltage is utilised to charge the battery. In this manner, energy may be saved in the battery, which can then be utilised to charge other

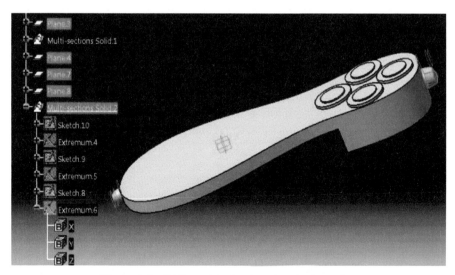

Figure 6.4 CAD model of a piezoelectric shoe sole.

devices. Material required for the present work is four piezoelectric plates, one battery (Li-polymer), four IN4007 Diodes, one capacitor, one switch and two LED's. Figure 6.4 illustrates the CAD model of the developed design.

6.5 Result and Discussion

Potassium sodium tartrate tetrahydrate, commonly called Rochelle salt, was synthesised and was used to generate energy via the piezoelectricity phenomena. The piezoelectric crystal has a diameter of around 2.5 cm and an output voltage of 1–2 V. The device displays the output voltage generated across capacitors from walking. The reaction force exerted on the ground while walking was calculated to be approximately 1–1.5 times the body weight. Using the assumption that an individual's body mass is roughly 80 kilograms on average, the force felt by the floor is of the order of 784 N. Using a tile displacement of 0.2 cm under an 800 N force results in a 4.16 ml/step that can be used.

Efficiency of piezoelectric material (η) is in general 2% as per eqn (6.1):

$$\eta = e_e/e_m, \tag{6.1}$$

$$e_m = F \times \text{displacement} \times 1.3 = 784\text{N} \times 0.02\text{cm} \times 1.3. \tag{6.2}$$

Therefore, e_m is 20.384 mV/step according to Equation (6.2) For the piezoelectric crystal, η is approximately 2%. Thus, 0.4076 mJ of energy is produced (e_e) per step. The current rating of the battery in use is 300 mAh, and the voltage rating of the battery is 4.2 V. Since, the steps required to fully charge the battery are 11128.5 steps (4536/0.0004076). Therefore, the distance to travel in order to fully charge is 5564.278 m (11128.5×0.5). Therefore, the present work is a low-cost alternative to demonstrating the use of piezoelectric sensors to capture electrical energy from human activities.

6.6 Conclusion and Future Scope

The humongous growth of the electronic industry in the past few decades has paved the way for the development of portable energy sources. There has also been raised awareness and concern about the environmental impacts of products. This has given an impetus to research portable energy-generating systems. Piezoelectric material in this regard is a great option, fitting very well into future technological needs. The present chapter discusses piezo-electricity in a generic sense in Section 6.1. In Section 6.2, a comprehensive literature survey is carried out on the various topics related to piezoelectricity and its different aspects. It briefly discusses the work of previous scientists in the mentioned domain and the importance of various parameters governing the techniques. On the basis of the literature review, the motivation for doing the present work and objectives related to the work are also summarised in Section 6.2. The materials and methodologies are discussed in Sections 6.3 and 6.4. It discusses the various raw materials used in the present study and the methodology adopted. Section 6.4 includes the results and discussion of the product developed in terms of the voltage generated to charge the battery of the phone in portable mode.

Let us take the Yamuna Expressway in Delhi, India as an example. The total budget of this project was Rs. 13,000 cr. If a 165 km stretch of piezoelectric road were laid additionally, the cost would be 13216 cr, which is a meagre 1.67% increase in the overall budget. While, each year, 44000 kWh of energy would have been generated from the same 1 kilometre single-lane road. So the total energy that can be produced from 165 km (165 km × 44,000 kWh) would be 7,260,000 kWh. Generally, the cost per unit of electricity in India is Rs. 5 per 1 kWh, so by calculation, the government could earn Rs. 3.63 cr.

The results obtained from the study are encouraging in terms of the functional outcomes of portable devices in the area of piezoelectricity. Our study also supports the previous evidence that piezoelectric devices in the era of clean and reliable resources will be of great utility. The concept may be expanded to provide electrical energy for larger load applications. It may also be used to charge portable devices in real time. Running, exercising and dancing would create more output than walking. Therefore, the sensors may be used in congested public places such as retail malls, airports, railway stations and so on to create and effectively utilise energy.

References

[1] Frank, R. (2005). *Handbook of modern sensors: physics, designs, and applications.* Anal Bioanal Chem **382,** 8–9. doi: 10.1007/s00216-004-3037-8.
[2] Howells, C. A. (2009) Piezoelectric energy harvesting, *Energy Conservation and Management,* 50(7), 1847-1850. doi: 10.1016/j.enconman.2009.02.020.

[3] Swallow, L. M., Luo, J. K., Siores, E., Patel, I., & Dodds, D. (2008). A piezoelectric fibre composite based energy harvesting device for potential wearable applications. *Smart Materials and Structures, 17*(2), 025017. doi: 10.1088/0964-1726/17/2/025017/meta.

[4] Covaci, C., & Gontean, A. (2020). Piezoelectric Energy Harvesting Solutions: A Review. *Sensors (Basel, Switzerland), 20*(12), 3512. doi: 10.3390/s201235122.

[5] Minazara, E., Vasic, D., Costa, F., & Poulin, G. (2006). Piezoelectric diaphragm for vibration energy harvesting. *Ultrasonics, 44 Suppl 1*, e699–e703. doi: 10.1016/j.ultras.2006.05.141.

[6] Zhao, J., & You, Z. (2014). A shoe-embedded piezoelectric energy harvester for wearable sensors. *Sensors, 14*(7), 12497–12510. doi: 10.3390/s140712497.

[7] Xue, H., Hu, Y., & Wang, Q. M. (2008). Broadband piezoelectric energy harvesting devices using multiple bimorphs with different operating frequencies. *IEEE transactions on ultrasonics, ferroelectrics, and frequency control, 55*(9), 2104–2108. doi: 10.1109/tuffc.903.

[8] Umeda, M., Nakamura, K. and Ucha, S. (1996). Analysis of Transformation of Mechanical Impact Energy to Electrical Energy Using a Piezoelectric Vibrator. *Japanese Journal of Applied Physics*, 35(1), 3267-3273. doi: 10.1143/JJAP.35.3267.

[9] Granstrom J., Feenstra J., Sodano H. A., Farinholt K. (2007). Energy harvesting from a backpack instrumented with piezoelectric shoulder straps. *Smart Mater. Struct.*, 16(5), 1810–1820. doi: 10.1088/0964-1726/16/5/036.

[10] Yang B., Kwang-Seok Y. (2012). Piezoelectric shell structures as wearable energy harvesters for effective power generation at low-frequency movement. *Sens. Actuators A Phys.*, 188, 427–433. doi: 10.1016/j.sna.2012.03.026

[11] Choi, W. B. (2001) *Potassium Sodium Tartrate* , John Wiley & Sons.

[12] Egbe, K-J. I., Nazar, A. M., Jiao, P., Yang, Y., Ye, X., Wang, H., (2021). Vibrational turbine piezoelectric nanogenerators for energy harvesting in multiphase flow fields. *Energy Reports*, 6384-6393. doi: 10.1016/j.egyr.2021.09.085.

[13] Kenji Uchino. (2010). *Advanced Piezoelectric Materials*. Woodhead Publishing Series in Electronic and Optical Materials.

[14] Kymissis, J., Kendall, C., Paradiso, J., & Gershenfeld, N. (1998, October). Parasitic power harvesting in shoes. In *Digest of papers. Second international symposium on wearable computers (Cat. No. 98EX215)* (pp. 132-139). IEEE. doi: 10.1109/ISWC.1998.7295399.

[15] Sodano, H. A., Inman, D. J. and Park, G. (2004) A Review of Power Harvesting from Vibration Using Piezoelectric Materials. The Shock and Vibration Digest, 36, 197-205. doi: 10.1177/0583102404043275

[16] Jo, S. H., Yoon, H., Shin, Y. C., Youn, B. D., (2021). An analytical model of a phononic crystal with a piezoelectric defect for energy harvesting using an electro elastically coupled transfer matrix. *International Journal of Mechanical Sciences,* 193, 106160. doi: 10.1016/j.ijmecsci.2020.106160.

[17] Shao, H., Chen, G., He, H., (2021). Elastic wave localization and energy harvesting defined by piezoelectric patches on phononic crystal waveguide. *Physics Letters A*, 403. doi: 10.1016/j.physleta.2021.127366.

[18] Jo, S. H., Yoon, H., Shin, Y. C., Youn, B. D., (2021). An analytical model of a phononic crystal with a piezoelectric defect for energy harvesting using an electroelastically coupled transfer matrix. *International Journal of Mechanical Sciences,* 193, 106160. doi: 10.1016/j.ijmecsci.2020.106160.

[19] Shao, H., Chen, G., He, H., (2021). Elastic wave localization and energy harvesting defined by piezoelectric patches on phononic crystal waveguide. *Physics Letters A*, 403. doi: 10.1016/j.physleta.2021.127366

[20] Zaarour, B., Zhu, L., Huang, C., Jin, X., Alghafari, H., Fang, J., & Lin, T. (2021). A review on piezoelectric fibers and nanowires for energy harvesting. *Journal of Industrial Textiles*, 51(2), 297–340. doi: 10.1177/1528083719870197

[21] Lu, Z. Q., Chen, J., Ding, H., Chen, L. Q., (2022). Energy harvesting of a fluid-conveying piezoelectric pipe. *Applied Mathematical Modelling*, 107, 165-181. doi: 10.1016/j.apm.2022.02.027.

[22] Starner, T., & Paradiso, J. A. (2004). Human generated power for mobile electronics. *Low-power electronics design*, 45, 1-35.

[23] Ishida, K., Huang, T. C., Honda, K., Shinozuka, Y., Fuketa, H., Yokota, T & Sakurai, T. (2012). Insole pedometer with piezoelectric energy harvester and 2 V organic circuits. *IEEE Journal of Solid-State Circuits*, 48(1), 255-264. doi: 10.1109/JSSC.2012.2221253.

7

Overview of Swarm Intelligence Techniques for Harvesting Solar Energy

Wei Hong Lim[1] and Abhishek Sharma[2]

[1]Faculty of Engineering, Technology and Built Environment,
UCSI University, Malaysia
[2]Department of Computer Science and Engineering,
Graphic Era Deemed to be University, Dehradun, India
E-mail: abhishek15491@gmail.com; limwh@ucsiuniversity.edu.my

Abstract

Photovoltaic systems are becoming increasingly popular in the energy production business. In spite of the advantages, photovoltaic (PV) systems have four major downsides: limited conversion precision, interrupted power supply, elevated manufacturing costs, and discontinuities of PV system output power. Numerous optimisation and control strategies have been suggested to address these issues. Numerous authors, however, trusted on traditional techniques that were based on instinctive, numerical, or analytical procedures. More effective optimisation approaches would improve the effectiveness of PV systems while lowering the price of energy produced. In this chapter, we will look at how Swarm Intelligence (SI) techniques can benefit PV systems. Specific attention is dedicated to two key areas: (1) active response and power quality enrichment of AC microgrids and (2) global maximum power point tracking.

Keywords: Photovoltaics; Swarm Intelligence; Parameter Extraction; MPPT; AC Microgrids.

7.1 Introduction

According to Richard Smalley, a famous scholar, energy is undoubtedly the most significant challenge confronting civilisation nowadays [1, 2]. Not only may solar energy help to democratise energy, but it also has the potential to greatly advance people's lives all around the world. The Sun is an immense cause of energy and plays a crucial part in many countries' energy production mix. Photovoltaic (PV) technology, in particular, is a fully grown, demonstrated, and dependable method of changing the Sun's massive energy into electric power. The advantage of PV technology is that it is customisable and robust. As a result, it can be speedily assembled in a variety of places. These places can range from traditional ground installations to residential and business structures. However, the significant cost of solar electricity in evaluation to other energy sources is one of the challenges to the widespread use of this approach. To drive down the rate of solar electricity and build innovative solar cells that can create more energy per unit area, advancement in the creation of novel materials and solar cell designs is critical. However, due to enhancements in computation capability and speed, SI is now evolving as just another efficient approach for assisting in the achievement of these goals. In three major areas: (1) active response and power quality enrichment AC microgrids and (2) global maximum power point tracking (GMPPT), we will look at how SI can be utilised in the domain of PV.

Examples of SI approaches include particle swarm optimisation (PSO) [3, 4], hill climbing (HC) [5], ant colony optimisation (ACO) [6, 7], grey wolf optimisation (GWO) [8, 9], etc. As a result, the purpose of this chapter is to deliver an impression of different AI methodologies and reveal how some of them might be implemented to raise the performance of PV systems in the aforementioned disciplines.

7.2 Brief Introduction to Swarm Intelligence Technique

Gerardo Beni and Jing Wang introduced Swarm Intelligence (SI) in 1989 [10, 11]. SI basically means combining the information of collaborative objects (people, creatures, etc.) to acquire the best solution to a given optimisation problem. A 'swarm' is a group of objects (people, insects, etc.). In other words, if we offer a problem statement to a solitary individual and train him or her to go through the problem and then provide an answer, we will only recognise the solution provided by that individual. The dilemma is that the solution provided by that person may not be the finest solution or may be destructive to others. To avoid this, we give the problem to a group of people (swarm) and ask them to acquire the appropriate solution possible for that problem, and then we compute all the responses together to discover the finest solution possible, so here we are utilising the information of the cluster as a whole to discover the finest solution or optimised solution for that problem, and that solution will be useful for all of them separately as well, so that is the concept beside swarm.

7.3 Advantages of Swarm Intelligence Techniques

The following are the primary benefits of SI-based systems:

(1) Collective robustness
   SI systems are stable in their essential working framework since they continue to operate collaboratively without any arbitrary supervision and no discrete exploring agent is required for the swarm to resume. SI systems, because of the aforementioned capability, can tolerate the downsides of a single search agent before causing significant harm to the swarm population [12].

(2) Scalability
   SI systems are vastly flexible. This is due to their method of control is not heavily reliant on swarm volume [13].

(3) Adaptability
   Because of inherent autonomy and auto-configuration functionality, SI systems can adapt to rapidly changing operational situations. As a result, such systems can adaptively change their actions on the fly based on the external environment, providing significant flexibility [14].

Furthermore, the following are some of the most imperative characteristics of a good optimisation algorithm:

- After each iterative process, the algorithm should maintain the finest solution and allocate it to the globally optimal variable.
- It must be able to save the previous solution so that the best option found thus far is not lost even if the entire populace depreciates.
- The method should proficiently find the search region at the beginning and exploit it at the final stage to obtain an optimum answer.
- The algorithm should be able to avert becoming stuck in local optima.
- The algorithm's parameters must be extremely adaptive in order for the algorithm to explore and utilise the search space at the initial stage

and conclusion of the process, respectively. The algorithm's working framework should be simple.

The benefits of an algorithm discussed above make it potentially capable of solving optimisation problems efficiently. Figure 7.1 depicts a simplistic step of the SI optimisation approaches to assist you in comprehending how optimisation algorithms perform.

The parameters of the chosen optimisation algorithm are described in advance in most SI optimisation techniques. These parameters often include algorithm constant values as well as other optimisation-related factors like the number of iterations, swarm sizes (number of particles), number of parameters to be optimised, and so on, which are required for the optimisation program code to execute smoothly. Following that, an objective function is formulated in the first step, which must be minimised or maximised in accordance with the implemented restrictions. During each iteration, the optimisation algorithm assesses and upgrades the objective function of each particle and governs the most suitable solution among all reasonable solutions. Based on the results, the finest particle grouping that fulfils the declared restrictions and objective function is preferred. Finally, the iterative progression comes to an end when the algorithm reaches the maximum number of iterations or an appropriate fitness value predetermined by the developer.

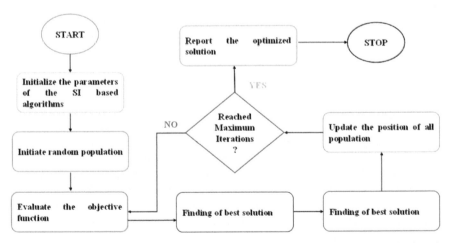

Figure 7.1 A simplified process diagram of swarm intelligence methods for finding the optimised solution.

7.4 Swarm Intelligence in Harvesting Solar Energy

In this chapter, we will reveal how SI has been effectively implemented in three distinct photovoltaic usages. A review of traditional and AI techniques will be introduced for each of these applications. Parameter extraction is the initial stage in PV system assessment and modelling. As a result, this section begins with the use of SI in the power reliability improvement of AC microgrids.

7.4.1 Active response and power quality enrichment of AC microgrids

Microgrids (MGs) have become increasingly smart, dispersed and adaptable. The electrical grids are being dominated by improved power devices and SI techniques, and this inclination is expected to continue for many years [15]. The increasing utilisation of innovative SI techniques in MG controls guarantees smooth incorporation and disassociation of DGs, improved power quality for end users, and improved power system transient stability. As a result, MGs are starting to emerge as a dependable and environmentally friendly origin of energy.

An MG is essentially a collection of loads and mini sources like wind turbines, fuel cells [16], solar PV [17] and micro-turbines that work together to form a single management framework that provides both energy and heat to the surrounding area [18]. The connectivity of MG with the utility grid typically necessitates the use of a non-linear power electronic gadget, like pulse width, modulated voltage source inverters (PWM-VSI) or converters. These electrical devices are crucial for integrating distributed energy resources (DER) into the power system and managing power transmission between DGs and the electric grid [19]. The main problem with these systems is that they generate non-linearity in both voltage and current as a result of the generation of elevated switching pulses, which obscures the system's power performance [20].

The serious issues affecting the power efficiency of MGs can be defined by their operation approach. The frequencies and voltage profiles of the MGs must be created by their control in islanded mode; or else, the system will malfunction due to the responsivity of the associated DG units and the absence of assistance from the core grid [21].

Furthermore, harmonic distortion of output power waveforms is a serious issue that is frequently induced by the rapid procedure of switching devices.

The long transient period can influence all system equipment, whether it happens during DG islanding or a sudden load alteration [22, 23].

These objectives are now excellently achieved by implementing SI control schemes, thanks to advancements in the domain of soft computational approaches. SI is a sort of artificial intelligence or machine intelligence that considerably decreases human labour [24]. According to the literature, SI methodologies surpass traditional methodologies in aspects of robustness, reliability, and fast response. However, traditional methodologies require memory to complete the aforesaid tasks. As a result, the addition of some extra capabilities in SI controllers helps make them more costly than traditional methodologies.

7.4.2 Advanced SI-based optimisation algorithm for improving AC microgrids power quality

This section gives an overview of advanced SI optimisation approaches for improving the active response and power performance of AC Microgrids.

The hybrid big bang-big crunch (HBB-BC) algorithm-based controller was developed to monitor the MG voltage stability [25]. For the same framework setup and optimisation problem, the achieved results were evaluated by comparing them to those of the PSO and big bang-big crunch (BB-BC) algorithms. However, because the investigated system did not use frequency control, the frequency settled to a lesser value (59.7 Hz) than the reference after a slight load alteration. Moreover, the feed-forward gains and droop coefficients were chosen provisionally, which does not guarantee the optimal assortment of the stated parameters. Dasgupta et al. [26] developed a spatial iterative learning controller (ILC) for grid voltage regulation throughout a grid-connected PV system. The recommended controller was used to keep the needed voltage at the load side while also removing grid harmonics. The artificial bee colony (ABC) algorithm was employed in a further experiment [56] to adjust the PI parameters for governing active and reactive power in an AC MG. Droop control was used to control voltage and frequency variations in both islanded and grid-connected systems. Besides that, the power and current controller variables of the studied MG system were optimised utilising the ABC optimisation method to achieve the best dynamic response [27].

For the optimised tuning of PI controllers in grid-connected hybrid wind/PV generating units, the firefly optimisation algorithm (FOA) was modelled and implemented. The study's goal was to adjust the voltage at the PCC and the system frequency under both load shifting and fault circumstances

Table 7.1 Advanced SI optimisation algorithms for improving AC microgrids power quality.

SI-based algorithms	Operating mode of microgrid	Operating conditions
GOA [31]	Islanded	Variations in load
FOA [32]	Islanded	Variations in load
MFO [33]	Islanded	Variations in load and source
Krill Herd algorithm [34]	Islanded	Variations in load and source
SSA [35]	Grid connected	Variations in source
Improved BFO [36]	Grid connected	Variations in source
ABC [27]	Islanded and grid connected	Variations in load

[28]. Banerjee et al. [29] created a seeker optimisation algorithm (SOA)-based controller to govern the reactive power of an MG powered by wind and diesel generating units. The investigators expanded on their investigation and suggested an online reactive power remittance procedure for islanded MG powered by wind and diesel generators [30]. Table 7.1 lists some of the most notable studies in this field that have been conducted by researchers from around the world.

7.5 Global Maximum Power Point Tracking

When compared to traditional energy sources such as oil, natural gas, and fossil fuels, solar power system is regarded as one of the most promising renewable energy sources due to its cleanliness, ample supply, and environmentally friendly nature. The complicated relationship between power output and PV input parameters leads to insufficient power extraction [37]. Maximum power point tracking (MPPT) is becoming the research focus to boost the effectiveness of the solar power system and make sure that the process point is always at the maximum power point to mitigate the aforementioned limitation (MPP) [38]. Peak uniform conditions can be magnificently monitored without partial shading conditions (PSC) using traditional methods such as perturb and observe (P&O), HC, and incremental conductance (INC). However, under PSC, the power output from a solar power system generates multiple peaks, with one global MPP (GMPP) and numerous other local peaks, as illustrated in Figure 7.2, complicating the HC MPPT technique's pursuit for the true maximum.

As a result, MPPT develops into a procedure based on evolutionary, heuristic and meta-heuristic methods. Because traditional HC methods struggle to monitor global peaks under PSC and rapidly shifting solar irradiance, it

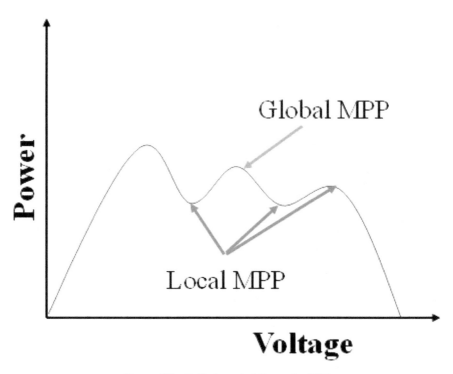

Figure 7.2   P–V characteristics under PSC.

is designed to track global peaks rather than local peaks [39]. Other methods to enhance solar energy's overall effectiveness, in addition to electronically implemented MPPTs, include integrated soft-computing weather prediction and adjusting the tilting angle of solar panels to trail the sun direction [40]. We only look at MPPT techniques based on SI for DC−DC converters in solar power systems. In general, all traditional MPPT techniques suffer from the same drawbacks, such as power oscillation, incapability to perform normally under PSC and quick irradiance alterations, capturing at one of the local MPPs and oscillation across the MPP [41–44]. Figure 7.3 depicts a typical MPPT schematic diagram, where PWM is identified as pulse width modulation.

7.5.1 Advanced SI-based optimisation algorithm for GMPPT

The authors suggested a customised form of PSO with both static and adaptive heuristic entities [45]. This amendment enhanced the inherent randomness of the classic PSO algorithm, resulting in a faster global maximum

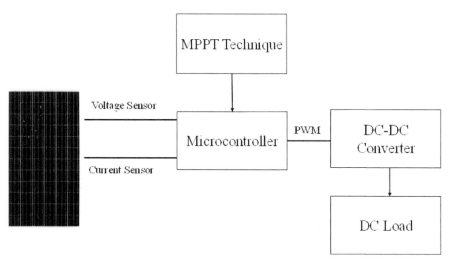

Figure 7.3 Schematic diagram of MPPT.

finding velocity. This study, however, lacked a correlation with advanced MPPT techniques. In another study [46], the authors used the ant colony optimisation (ACO) method to monitor the GMPP in PSCs. However, this study appears to lack an innovative analysis of the MPPT controller under a variety of conditions.

In [47], moth flame optimisation (MFO) is used as a unique method for optimal PV source exploitation under PSC. In aspects of tracking capacity, effectiveness and steady state efficiency, it outperforms INC and PSO. Another hybrid MPPT is the ANN-CGSVM technique, which merges two prevailing machine learning (ML) techniques, coarse-Gaussian support vector machine (CGSVM) and ANN. CGSVM is a data mining technique that is a type of non-linear SVM learning technique [48]. An improved version of the gravitational search algorithm (IGSA) is proposed by the authors in [49] for tracking MPP under varying irradiance levels. Simulation results show that IGSA outperforms the GSA and PSO in aspects of less oscillation around the GMPP.

Mohanty et al. [50] conducted a thorough comparison of the P&O, IPSO, and GWO procedures. Hardware execution is carried out with the dSPACE1104 integrating platform and the TMS320F240 DSP, and a hall effect sensor is employed to detect both voltage and current and give real-time data to the controller. Mohanty et al. [51] suggested a hybrid GWO-aided P&O technique that addresses the difficult challenges of ambiguous hunt spaces. GWO is shown to be the best calculation due to its ease of use, adaptability, individual system and evasion of local optima. Authors in [52] presented the ABC-based MPPT strategy and the outcomes of ABC were evaluated against the PSO and improved P&O (IPO) approaches. The ABC method is nearly 99.99% effective. The tracking speeds are impressive, as they are 4.234, 9.375 and 1.425 s for the ABC, PSO, and IPO techniques. Though, when the shading patterns are altered instantly, the efficiencies decrease. Padmanabhan et al. [53] recently developed the best hybrid ANFIS-ABC technique to date. This approach is implemented in hardware on the DSP platform. The utilised PV system is grid-integrated, and the efficacy is 98.39% while maintaining grid voltage and current at 2.3% and 2.5%, respectively. The authors in [54] anticipated the ABC approach and compared this approach to the other three in terms of monitoring speed and effectiveness.

Among other optimising estimations, the CS method has been regarded as unique [55]. The central concept of CS is brood parasitism, which involves laying eggs inside the homes of other birds and is classified as integral component, nest invasion and collaboration. Deb et al. [56] provide a thorough explanation of the multi-objective layout of the CS optimisation process. For simplicity in depicting the CS approach, the following three rules are currently used: (i) each cuckoo lays each egg in turn, and dumps its egg in an arbitrarily chosen home; (ii) the best quality of eggs within that nest will persevere to the following generations; and (iii) the volume of obtainable host homes is static, and the egg laid by a cuckoo is discovered by the host bird with a possibility of [0, 1]. In this case, the host bird can either reject the egg or abandon the home and build a new one. This decisive assumption can be estimated for simplicity by the part dad of the n homes being replaced by new homes (with new irregular arrangements). When the bird realises the egg is not his own, it will decimate it with levy flight.

7.6 Conclusion

This chapter examined a number of MPPT techniques for raising the maximum power output of a PV system under PSC conditions. A comprehensive review of the literature for numerous MPPT approaches is conferred, with specific parameters taken into account. Furthermore, it is deduced from the literature that using MPPT controllers is the best method for monitoring the MPP under PSC's, which has paved the way for further research. Based on this review, advanced SI approaches with enhanced exploration and exploitation balancing capabilities are expected to be effective in improving the power performance and dynamic response of microgrid systems in both isolated and grid-tied modes of operation. The competence of these optimisation methods is highly sought after in light of the quickly growing invasion of renewable energy sources into the present structure, which is one of the prevailing era's quickest-rising classifications.

References

[1] Ghannam, R., P. V. Klaine, and M. Imran, Artificial intelligence for photovoltaic systems, in Solar Photovoltaic Power Plants. 2019, Springer. p. 121-142.
[2] Sharma, A., et al., Improved moth flame optimization algorithm based on opposition-based learning and Lévy flight distribution for parameter estimation of solar module. Energy Reports, 2022. 8: p. 6576-6592.
[3] Marini, F. and B. Walczak, Particle swarm optimization (PSO). A tutorial. Chemometrics and Intelligent Laboratory Systems, 2015. 149: p. 153-165.
[4] Singh, A., et al., An Investigation on Hybrid Particle Swarm Optimization Algorithms for Parameter Optimization of PV Cells. Electronics, 2022. 11(6): p. 909.
[5] Liu, F., et al. Comparison of P&O and hill climbing MPPT methods for grid-connected PV converter. in 2008 3rd IEEE Conference on Industrial Electronics and Applications. 2008. IEEE.
[6] Dorigo, M., M. Birattari, and T. Stutzle, Ant colony optimization. IEEE computational intelligence magazine, 2006. 1(4): p. 28-39.
[7] Sharma, A., et al., A review on artificial bee colony and it's engineering applications. Journal of Critical Reviews, 2020. 7(11): p. 4097-4107.
[8] Faris, H., et al., Grey wolf optimizer: a review of recent variants and applications. Neural computing and applications, 2018. 30(2): p. 413-435.
[9] Sharma, A., et al., An effective method for parameter estimation of solar PV cell using Grey-wolf optimization technique. International Journal of Mathematical, Engineering and Management Sciences, 2021. 6(3): p. 911.
[10] Beni, G. and J. Wang, Swarm intelligence in cellular robotic systems, in Robots and biological systems: towards a new bionics? 1993, Springer. p. 703-712.
[11] Sharma, A., et al., Swarm intelligence: foundation, principles, and engineering applications. 2022: CRC Press.

[12] Dorigo, M., In The Editorial of the First Issue of: Swarm Intelligence Journal. LLC, 2007. 1(1).

[13] Valentinuzzi, M. E., Handbook of bioinspired algorithms and applications. BioMedical Engineering OnLine, 2006. 5: p. 47.

[14] Sharma, A., et al., Path planning for multiple targets interception by the swarm of UAVs based on swarm intelligence algorithms: A review. IETE Technical Review, 2021: p. 1-23.

[15] Guerrero, J. M., et al., Hierarchical control of droop-controlled AC and DC microgrids—A general approach toward standardization. IEEE Transactions on industrial electronics, 2010. 58(1): p. 158-172.

[16] Sharma, A., et al., A Novel Opposition-Based Arithmetic Optimization Algorithm for Parameter Extraction of PEM Fuel Cell. Electronics, 2021. 10(22): p. 2834.

[17] Pachauri, R. K., et al., Applied Soft Computing and Embedded System Applications in Solar Energy. 2021: CRC Press.

[18] Lasseter, R. H. Microgrids. in 2002 IEEE power engineering society winter meeting. Conference proceedings (Cat. No. 02CH37309). 2002. IEEE.

[19] Nejabatkhah, F. and Y. W. Li, Overview of power management strategies of hybrid AC/DC microgrid. IEEE Transactions on power electronics, 2014. 30(12): p. 7072-7089.

[20] Zeng, Z., et al., Topologies and control strategies of multi-functional grid-connected inverters for power quality enhancement: A comprehensive review. Renewable and Sustainable Energy Reviews, 2013. 24: p. 223-270.

[21] Pedrasa, M. A. and T. Spooner, A survey of techniques used to control microgrid generation and storage during island operation. AUPEC2006, 2006. 1: p. 15.

[22] Mohamed, Y., New control algorithms for the distributed generation interface in grid-connected and micro-grid systems. 2008.

[23] Majumder, R., et al., Improvement of stability and load sharing in an autonomous microgrid using supplementary droop control loop. IEEE transactions on power systems, 2009. 25(2): p. 796-808.

[24] Kow, K. W., et al., A review on performance of artificial intelligence and conventional method in mitigating PV grid-tied related power quality events. Renewable and Sustainable Energy Reviews, 2016. 56: p. 334-346.

[25] Sedighizadeh, M., M. Esmaili, and A. Eisapour-Moarref, Voltage and frequency regulation in autonomous microgrids using Hybrid Big Bang-Big Crunch algorithm. Applied Soft Computing, 2017. 52: p. 176-189.

[26] Dasgupta, S., S. Sahoo, and S. Panda. Design of a spatial iterative learning controller for single phase series connected PV module inverter for grid voltage compensation. in The 2010 International Power Electronics Conference-ECCE ASIA-. 2010. IEEE.

[27] Bai, W., M. R. Abedi, and K. Y. Lee, Distributed generation system control strategies with PV and fuel cell in microgrid operation. Control Engineering Practice, 2016. 53: p. 184-193.

[28] Chaurasia, G. S., et al., A meta-heuristic firefly algorithm based smart control strategy and analysis of a grid connected hybrid photovoltaic/wind distributed generation system. Solar Energy, 2017. 150: p. 265-274.

[29] Banerjee, A., V. Mukherjee, and S. Ghoshal, Modeling and seeker optimization based simulation for intelligent reactive power control of an isolated hybrid power system. Swarm and Evolutionary Computation, 2013. 13: p. 85-100.

[30] Banerjee, A., V. Mukherjee, and S. Ghoshal, Intelligent fuzzy-based reactive power compensation of an isolated hybrid power system. International Journal of Electrical Power & Energy Systems, 2014. 57: p. 164-177.

[31] Jumani, T. A., et al., Optimal voltage and frequency control of an islanded microgrid using grasshopper optimization algorithm. Energies, 2018. 11(11): p. 3191.

[32] Akbari, M., M. Golkar, and S. M. Tafreshi. Firefly algorithm-based voltage and frequency control of a hybrid ac-dc microgrid. in 2012 Proceedings of 17th Conference on Electrical Power Distribution. 2012. IEEE.

[33] Houshyari, M., H. Shayeghi, and A. Younesi. Comparison of PID type controller performance in microgrid frequency deviation enhancement using MFO algorithm. in Univ. of Mohaghegh Ardabili. 2016.

[34] Shayanfar, H., H. Shayeghi, and A. Younesi. Optimal PID controller design using Krill Herd algorithm for frequency stabilizing in an isolated wind-diesel system. in proceedings on the international conference on artificial intelligence (ICAI). 2015. The Steering Committee of The World Congress in Computer Science, Computer ....

[35] Jumani, T. A., et al., Salp swarm optimization algorithm-based controller for dynamic response and power quality enhancement of an islanded microgrid. Processes, 2019. 7(11): p. 840.

[36] Saad, N. H., A. A. El-Sattar, and M. E. Marei, Improved bacterial foraging optimization for grid connected wind energy conversion system based PMSG with matrix converter. Ain Shams Engineering Journal, 2018. 9(4): p. 2183-2193.

[37] Yang, Y. and H. Wen, Adaptive perturb and observe maximum power point tracking with current predictive and decoupled power control for grid-connected photovoltaic inverters. Journal of Modern Power Systems and Clean Energy, 2019. 7(2): p. 422-432.

[38] Yung, Y. K., et al., Central composite design (CCD) for parameters optimization of maximum power point tracking (MPPT) by response surface methodology (RSM). Journal of Mechanics of Continua and Mathematical Sciences, 2019. 1: p. 259.

[39] Sharma, A., et al., A Novel TSA-PSO Based Hybrid Algorithm for GMPP Tracking under Partial Shading Conditions. Energies, 2022. 15(9): p. 3164.

[40] Mishra, D. P. and S. Chakraborty. Application of soft computing in simulation of solar power tracking. in 2018 Technologies for Smart-City Energy Security and Power (ICSESP). 2018. IEEE.

[41] Jately, V., et al., Experimental analysis of hill-climbing MPPT algorithms under low irradiance levels. Renewable and Sustainable Energy Reviews, 2021. 150: p. 111467.

[42] Sharma, A., et al., An effective method for parameter estimation of a solar cell. Electronics, 2021. 10(3): p. 312.

[43] Rajput, S., et al., An approval of MPPT based on PV Cell's simplified equivalent circuit during fast-shading conditions. Electronics, 2019. 8(9): p. 1060.44. Rajput, S., M. Averbukh, and N. Rodriguez, Energy Harvesting and Energy Storage Systems. 2022, MDPI. p. 984.

[44] Li, H., et al., An overall distribution particle swarm optimization MPPT algorithm for photovoltaic system under partial shading. IEEE Transactions on Industrial Electronics, 2018. 66(1): p. 265-275.

[45] Jiang, L. L., D. L. Maskell, and J. C. Patra, A novel ant colony optimization-based maximum power point tracking for photovoltaic systems under partially shaded conditions. Energy and Buildings, 2013. 58: p. 227-236.

[46] Aouchiche, N., et al., AI-based global MPPT for partial shaded grid connected PV plant via MFO approach. Solar Energy, 2018. 171: p. 593-603.

[47] Farayola, A. M., A. N. Hasan, and A. Ali, Efficient photovoltaic MPPT system using coarse gaussian support vector machine and artificial neural network techniques. International Journal of Innovative Computing Information and Control (IJICIC), 2018. 14(1): p. 323-329.

[48] Li, L.-L., et al., A maximum power point tracking method for PV system with improved gravitational search algorithm. Applied Soft Computing, 2018. 65: p. 333-348.

[49] Mohanty, S., B. Subudhi, and P. K. Ray, A new MPPT design using grey wolf optimization technique for photovoltaic system under partial shading conditions. IEEE Transactions on Sustainable Energy, 2015. 7(1): p. 181-188.

[50] Mohanty, S., B. Subudhi, and P. K. Ray, A grey wolf-assisted perturb & observe MPPT algorithm for a PV system. IEEE Transactions on Energy Conversion, 2016. 32(1): p. 340-347.

[51] Sundareswaran, K., et al., Enhanced energy output from a PV system under partial shaded conditions through artificial bee colony. IEEE transactions on sustainable energy, 2014. 6(1): p. 198-209.

[52] Padmanaban, S., et al., A hybrid ANFIS-ABC based MPPT controller for PV system with anti-islanding grid protection: Experimental realization. Ieee Access, 2019. 7: p. 103377-103389.

[53] Li, N., et al. Maximum power point tracking control based on modified ABC algorithm for shaded PV system. in 2019 AEIT International Conference of Electrical and Electronic Technologies for Automotive (AEIT AUTOMOTIVE). 2019. IEEE.

[54] Sharma, A., et al., Cuckoo search algorithm: A review of recent variants and engineering applications. Metaheuristic and Evolutionary Computation: Algorithms and Applications, 2021: p. 177-194.

[55] Yang, X.-S. and S. Deb, Multiobjective cuckoo search for design optimization. Computers & Operations Research, 2013. 40(6): p. 1616-1624.

8

Energy Harvesting from Conducting Nanocomposites

Ankit Kumar Srivastava[1], Swati Saxena[2], Sonika,[3] and Sushil Kumar Verma[4]

[1]Department of Physics, Indrashil University, India
[2]Department of Physics, Sardar Vallabhbhai National Institute of Technology, India
[3]Department of Physics, Rajiv Gandhi University, Doimukh, Itanagar, India
[4]Department of Chemical Engineering, Indian Institute of Technology, Guwahati, India
E-mail: pushpankit@gmail.com; swastisaxenaa@gmail.com; sonika.gupta@rgu.ac.in; sushilnano@gmail.com; swastisaxenaa@gmail.com

Abstract

Contemporary electronic gadgets consume minimum energy and may be charged by renewable energy sources. Because of their great availability in technological situations, composite materials are appealing sources for energy harvesting. Piezoelectric transmission is one of the many methods for converting mechanical energy towards electrical energy. It has maximum output energy at micro energy levels. The quantity of electrical energy gathered by piezoelectric energy harvesters is exactly proportional to the strain experienced by the transducer. Mostly piezoelectric transducers are built of fragile polycrystalline ceramic materials like PZT. This reduces the harvester's maximum permissible strain and, as a result, the amount of electricity gathered. Electroactive polymers, because of their flexibility, are a feasible choice in certain situations. Energy harvesting from polymer electrolytes is the subject of this chapter. Polymers like polyurethane polymers are widely used in energy harvesting systems because of their adaptability, low cost and good

177

electro-mechanical coupling skills. The development of energy harvesters that use these materials is then explained. Harvesters made of graphene and carbon nanotubes are defined as conducting polymer composites. To have a better perspective of energy harvesting from conductive polymers, a comparison of the harvesting capacities of different conductive polymers and the obstacles they encounter is explored.

Keywords: Energy Harvesting; Nanocomposite; Conducting Polymers; Carbon Nanotubes; Graphene.

8.1 Introduction and Background

The globe faces a fundamental issue at the start of the twenty-first century: developing energy sources that can fulfil ever-increasing energy demand while still being environmentally friendly. The world's energy needs are constantly increasing, owing to population increase and rapid economic development. During the producing power, about 60% of the electrical energy gathered from power plants for use in our homes is lost as waste heat [1], and somewhere between 8% and 15% is dissipated as heat in the electric cables for its transit and transformation [2]. As a result, just 35% of the energy generated in a power plant reaches our homes. Another example is the efficiency achieved in transit, in which 40% of the power generated in a car is discarded as heat, and another 30% is used to cool the engine, giving rise to a total of 70% wasted energy, even without accounting for the CO_2 emissions into the environment caused by the additional 70% of the gasoline that must be used. In other words, in everyday living and industrial output, a large amount of low-quality thermal energy is necessarily created and, tragically, squandered [3]. Thermoelectric functional materials are renewable materials that may convert electricity and heat, making use of solid internal carriers' mobility, even at extremely small temperature differences from room temperature. Thermal energy (TE) devices, in comparison to other typical new energy technologies, offer several unique characteristics, such as the absence of moving components and noise, and a long operational lifetime, and have therefore emerged as a highly viable option to replace existing energy materials [4–6]. Thermoelectrical materials are widely employed in the army, aircraft and other high technological domains, as well as in medicinal thermostats, microsensors, and other civilian applications [7, 8]. It is in this context that thermoelectric generators can assist in a more sustainable environment by utilising their capacity to transform temperature changes into electric energy, that is, getting electricity from squandered heat [10]. According to the script of EIA, United States Energy Department, the worldwide energy demand is rising by 49%. (or 1.4% per year), which is expected to be 739 quadrillion BTU in 2035. These calculations are the reason for concern, not just because supplying power to these measurements will be difficult, but because of the bulk of power (85%) is now provided by the use of fossil fuels. Fossil fuel burning emits a substantial atmospheric CO_2, a recognised greenhouse gas that contributes to climate change. Carbon-free renewable sources must be developed on a large scale in the near future to avert a climate change disaster by stabilising CO_2 levels at appropriate goal levels.

8.2 Energy Harvesting

Energy harvesting was previously a non-issue for developers, due to the introduction of ultra-low-energy MCUs. However, because batteries research was unable to keep up with smaller energy harvesting started receiving serious consideration because to the proliferation of portable electronics. For example, wireless sensor networks would not be possible without ultra-low-energy MCUs, which are aided in addition by micropower harvesting units [11–12].

Solar, thermal, RF and piezoelectric energy sources are employed in the most common energy harvesting systems.

1. Solar cells or photovoltaic (PV) panels transform light energy into electricity. Among the numerous energy-collecting devices, photovoltaic cells have the maximum power density and production.
2. Heat is converted into electricity using thermoelectric energy harvesters. They are made up of thermocoupler arrays that create a voltage when the temperature difference between their bi-metal junction is detected (effect of Seebeck). The opposite is true: applying power to a thermo-couple (TC) junction causes one junction to heat, however, the other one cools, and this is the way to work of heat pumps (effect of Peltier).
3. Radiofrequency harvesters collect radiofrequency radiation from the environment, correct it, amplify it and utilise it for ultra-low-energy electronics. Radiofrequency identification works on the same concept, but instead of capturing ambient RF, it reacts to a strong RF field aimed at the sensor.

4. Pressure or tension is converted into electricity using piezoelectric transducers. Motors, airfoils and roadbed vibrations are often piezoelectric energy harvesters, which are reported as anomalies.

Other energy collecting systems are being investigated, but the redisplay is the frontrunner. Such four energy harvesting businesses will continue to grow at a rapid pace for many years to come, thanks to the expansion of battery-powered transportable consumer, commercial and diagnostic supplies.

8.3 Energy Harvesting Sources

Energy harvesting resources are those found in the surrounding environment that has the ability to generate enough energy to power sensor networks in smart environments in full or in part. Energy harvesting sources may be divided into two classes based on their characteristics:

Natural sources are those that are easily available from the environment, such as solar power, wind energy and geothermal energy, whereas artificial sources are those that are produced by human or system actions. They are not a natural element of the ecosystem. Human motion, pressure on floor slabs inserts when walking or running and system vibrations while operations are all examples.

Table 8.1 lists several sources of energy for energy harvesting, as well as the kind of source and usual harvesting power. For two reasons, system designers must consider the sort of energy-collecting source. Natural sources are impacted by natural variables including weather, temperature and season, whereas artificial sources are influenced by human and machine systems' schedules and impacts. This will have an influence on the prediction mechanisms of each generation source, for example. Second, natural resources do not require additional energy to produce. Harvesting natural resources at a large scale might have environmental consequences, which are outside the scope of our research into micro-scale energy harvesting technologies. Artificial resources, on the other hand, necessitate the expenditure of energy by human/machine systems in order to create atmospheric harvestable energy. If the energy generated is mostly utilised for other reasons, such as illuminating a room or running a computer system, it should not be regarded as a cost. As a result, the available harvestable energy is only a by-product of this process. However, creating energy is considered a cost if it is primarily utilised to create harvestable energy. This can happen when a light bulb is left on for a

Table 8.1 Comparison in the study of various forms of energy consumption.

Serial no.	Sources of energy	Energy amount (TW)
1	Hydropower resources	≤ 0.50
2	Ocean and tide energy	≤ 2.00
3	Wind energy	$2.00-4.00$
4	Solar power	120000

few more hours to charge a sensor using photovoltaic power, or when a radio spectrum is created to charge an RFID sensor.

Thermoelectric materials are renewable energy sources that may convert heat to electricity in a direct manner. As a result, they may be used for energy collecting and local cooling in TE generators. Spent fuel, importantly small heat waste, might be beneficial. Currently, organic conductive polymers carbon nanocomposites are used as thermoelectric composites and have gotten a lot of interest because of their synergistic effects, which elaborate the benefits of both carbon nanoparticles and polymers.

Thermoelectric generators are being called upon to help enhance the efficiency of the real energy system by recovering today's lost heat in the rush to find alternative energy sources. These features of thermoelectric nanostructuring, ranging from so-called 3D nano bulk materials to the inclusion of 0D quantum dots in thermoelectric structures [61], are studied.

According to our comprehensive literature research, maintaining a homogeneous dispersion and long-term stability are critical for achieving good thermal characteristics of CNT nanofluids throughout time. The objective of maintaining the aforementioned parameters is undoubtedly tough due to the hydrophobicity of CNTs against most fluids and the strong van der Waals contact among CNT nanoparticles. Nonetheless, many scholars have sought to meet those criteria in a variety of ways. However, several problems must be discovered and addressed for various applications of CNT nanofluids, particularly for solar systems. Its commercialisation is hampered by two key factors: stability and manufacturing costs. As a result, in order to employ nanofluid in solar collectors, most collector designs must be restructured to fulfil the practical requirements for water heating systems used both domestically and industrially. Nanofluids are projected to have a significant influence not only on industries and technical sectors but also on improving the quality of human life, if these hurdles are overcome. Meanwhile, more recent research [13–15] has demonstrated the feasibility of employing hybrid/composite and magnetic nanofluid to improve heat transmission.

Despite the fact that none have been employed in solar energy systems, further scientific study is needed to assure the highest level of performance of nanofluid in solar thermal engineering devices. A group of scientists has demonstrated that energy may be harvested from carbon nanotube yarns. Researchers from the University of Texas in Dallas led the team. For more than 10 years, experts there have been working on producing carbon nanotube-based yarns. They have finally succeeded in turning carbon nanotube threads into energy-harvesting devices by twisting and extending them.

The first findings suggest that these nanotubes might be utilised to power tiny sensor nodes in IoT applications right away. Scientists believe that by stretching and flexing in response to the motions of waves crashing, nanotube yarns may generate large quantities of energy. The piezoelectric effect appears to be used by the nanotube.

8.4 Energy Harvesting Storage

1. Micropower energy is frequently intermittent, but even if it is not, because the output is usually so low, a boost valve is necessary to keep the accumulated energy in check.
2. In space-constrained portable applications, small rechargeable Li-ion batteries is common.
3. Current spikes, for example when a sensor shows burst data, are difficult for micropower energy devices to handle. Energy harvesters often need a big supercapacitor/capacitor to buffer abrupt surges in the stipulation, in addition to requiring power control.
4. Due to the exceptionally close closeness of its conducting layers, electrical double layer capacitor is also recognised as supercapacitors. The energy density of these supercapacitors is lower than that of batteries, but the power density is much higher. They may be depleted in a matter of seconds, unlike batteries, making them ideally adapted to handle abrupt increases it is demand. While they have minimum internal resistance than thin-film batteries, allowing them to drain more slowly over time, hence they are frequently employed combined in ultra-low-power applications.

8.5 Energy Collection from Conducting Nanocomposites Development Tools

8.5.1 Development tool for thermal energy harvesting

Energy harvesting encompasses a wide range of continually changing technologies. Development tools are becoming increasingly important as design complexity rises and design cycles shorten. Development kits and boards in the field of energy harvesting allow designers to assess and have been familiar with the newest harvesting power technologies and products.

Beyond circuit modelling, the iterative design process must incorporate development and testing. Engineers are regularly forced to use the ostensibly unavoidable breadboard. With low set-up time, development kits provide a quick way to get to the heart of development.

Development tools can also give advantages for immediate applicable, tested circuits, widely obtainable printed circuit layouts and a similar level from which to generate designs, in addition to a faster time to market. Energy harvesting encompasses a wide range of continually changing technologies. Development tools are becoming increasingly important as design complexity rises and design cycles shorten. Development kits and boards in the field of energy harvesting allow designers to assess and make them familiar with the newest harvesting energy technologies and products.

8.5.2 A quick look at carbon nanotubes (CNT) and graphene

Carbon traditionally has been an extremely versatile and all-purpose element, with a wide range of allotropes that have several uses in the materials world. Many various types of carbon forms are in existence, but carbon nanotubes and graphene have emerged as the most intriguing materials, with limitless research potential in all aspects of science. Carbon nanotube always was a famous name in science since its discovery by the eminent scientist Iijima [16]. CNTs typically have lengths in the micrometre range. Depending on the lattice vectors and chiral angles, CNTs exhibit three distinct chirality: armchair, zigzag and chiral [17]. CNTs have become the most expected substance in society, with vast uses ranging from industrial to nanotechnology, including, energy storage, modular electronics, conductive polymers and structural composites. Another type of carbon that has just been investigated for diverse purposes is graphene. Graphene is a relatively new substance in the realm of materials, but due to its unique structure, it has demonstrated amazing mechanical, thermal and electrical capabilities. Graphene is a two-dimensional carbon allotrope composed of a hexagonal lattice of carbon atoms on a planar surface. Although graphene provides the foundation for other carbon nanostructures such as carbon nanotubes and fullerene. In 2004, graphene became a reality for the first time. Following that, a burst of interest in graphene's structure and characteristics erupted, and graphene did not let us down. Graphene's properties, including charge transport and tensile strength, were measured in detail and found to be quite high. With these unique features, graphene might be used in a variety of applications, thin and flexible screens, solar cells, electronics, medical, pharmaceutical and industrial activities are only a few of the topics covered [18–[23]]. CNTs and graphene have been widely researched in a range of thermoplastic and thermosetting polymer systems in order to improve electrical and mechanical properties. Several researchers have reported work employing polyurethane, polystyrene, polycarbonate, ABS, PMMA, polyethylene, epoxy and phenolic systems [24–33]. CNT and graphene have also been utilised with polymers to generate self-healing nanocomposites because of their excellent thermal conductivity and stability. A graphene structure, an SWCNT structure and an MWCNT structure [21]. The self-healing phenomenon accomplished with graphene, as well as the polymers afterwards utilising CNTs in a similar fashion, is briefly discussed in the following sections.

8.5.3 Self-healing polymer composites based on graphene

Intrinsic Defect (3.1) Graphene for Healing: Graphene has the ability to fix innate flaws. The healing effect is caused by the graphene structure being reconstructed (knitted) [35]. Carbon atoms from the exterior areas rush in to fill the gaps when a vacancy defect forms [36]. Impurities, such as hydrocarbon impurities, are typically the cause of these additional carbon atoms. Carbon atoms are transported from surrounding hydrocarbon impurities and fill the hole created when graphene is scraped in the presence of metals., resulting in the strange reknitting of graphene holes [37]. This is due to graphene's inherent self-healing characteristic, which opens up new options for graphene applications employing diverse processes such as the e-beam technique and the etching process. A molecular dynamics simulation investigation of a stiff C60 molecule causing nanodamage in a suspended graphene monolayer reveals an effective self-healing method with the right thermal treatment [38].

The self-healing mechanism is defined as a two-step process: (a) the creation of local curvature around the defects caused by the damage and (b) defect rebuilding resulting in noise filtering of the contour generated by the destruction, finally resulting in the destruction being reduced. Graphene's capacity to self-heal is mostly determined by two elements:

(1) the size of the damage done and
(2) the temperature variation.

As a result, graphene, being a single atom layer, is worth highlighting, may preserve the power of moving current and thus damage repair, in the same manner, graphene may readily repair damage in microstructures when combined with polymer composites. The self-healing in polymer composites using CNTs and graphene is the major emphasis of this chapter. Because of their flat structure and hence superior thermal contraction, high-performance thermal interface materials are another promising use of graphene-based polymer composites. Many researchers have found that adding graphene to a material improves heat conductivity. This will increase the polymer matrix's heat transport capabilities, resulting in improved self-healing in response to thermal or other external stimuli. Integrating graphene into polyurethane and epoxy are two examples of polymers. Many investigations on self-healing property analyses have been undertaken.

8.5.4 Carbon nanotube-based as self-healing polymer nanocomposites

Since their discovery, carbon nanotubes (CNTs) have been the subject of material study due to their ability to impart excellent mechanical qualities. CNT-reinforced materials offer a wide range of uses in a variety of industries. Apart from the aforementioned uses, In order to truly comprehend the self-healing characteristics of CNT-based polymer composites, CNT has been extensively studied. On one hand, it has a 1D tubular structure with strong reinforcing properties, while on the other, it has amazing thermal conductivity and heat transmission capabilities, CNT may be employed in self-healing materials in both extrinsic and intrinsic techniques.

8.5.5 Extrinsic self-healing polymers with CNTs

The capsule-based and fluid overload approaches of extrinsic self-healing can both be used. CNTs can be employed as a nanoreservoir for the healing agent because of their effective 1D nanotubular structure. Second, because of its structural enhancing function, it may be employed as a healing agent in fluid overload self-healing with other polymers. Finally, CNTs can be reinforced in microcapsule-based healable polymers to restore mechanical strength.

8.5.6 Carbon nanotubes as nanoreservoirs

Single-walled carbon nanotubes (SWCNTs) can be employed as nanoreservoirs of healing agents in microvascular self-healing approaches [39]. The mechanics of fluid flow from a ruptured SWCNT after the injury are studied

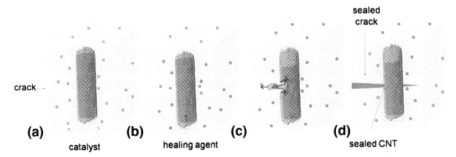

Figure 8.1   Carbon nanotubes act as nano reservoirs in a self-healing mechanism (courtesy: Deepalekshmi Ponnamma et al.).

in detail, with the fluid resembling a healing agent. According to the study of the dynamics, the employment of SWCNTs as a self-healing container reinforces the overall system mechanically. The following are the primary elements that influence the quantity of a healing agent that must be kept in the SWCNT reservoir for the self-healing process.

These criteria will be determined by the material's final uses when creating a realistic CNT-based self-healing system [39]. The self-sustained diffusion approach was employed to intercalate liquid monomers used as therapeutic agents into CNTs [40]. First, a semisolid solution of empty CNTs in benzene was combined with the healing agent (dicyclopentadiene (DCPD) or isophorone diisocyanate). After sonication, the benzene in the solution was evaporated, enabling the solutes to permeate into the CNTs. After the intercalation, fresh benzene was added and sonicated for 3–4 minutes to clean the exterior of the CNTs. If a more feasible approach for loading self-healing compounds into CNTs can be developed, mass production of smart composites with integrated self-healing agents might begin.

CNTs as a Reinforcing Filler in Healable Polymers Based on Capsules There are three primary drawbacks of different self-healing capsule and vascular-based polymer systems:

- Difficulty embedding capsules with healing agents within polymer systems.
- Capsule and vascular implantation cause material mechanical strength to deteriorate.

After a single crack, the mending agent is depleted. When compared to pure polymers, CNTs are desirable as reinforcing conductive fillers because of their enormous electrical conductivity and outstanding mechanical characteristics. As a consequence, CNT reinforcement can compensate for the loss of strength caused by capsule embedment. The healing agent was microencapsulated inside the polymer system, yielding an electrically conducting self-healing epoxy-based covering containing CNT fillers [41–44]. Although it is generally known that incorporating microcapsules into a polymer system diminishes mechanical strength, this flaw was solved by including SWCNTs in the polymer system. Adding SWCNTs to the system enhanced the elastic modulus and hardness of the samples substantially, according to nanoindentation measurements. As a result, CNTs may readily improve the physical characteristics of composites, which will not only broaden their applications but also assist restore mechanical qualities that have been lost owing to capsule embedment.

8.5.7 Carbon nanotubes as effective healing agents

CNTs have also been exploited as healing agents in vascular-based self-healing (SH) polymer. Nanocomposite including SWCNTs and 5 ethylidene-2 norbornene (5E-2N), which was mixed with ruthenium Grubb's catalyst [45], was entered into epoxy resin system's empty channels to be employed like a healing-agent.

For example, a micro-vascular SH method: To evaluate the self-healing behaviour, a mass was applied to raise the impact hole. The damaged sample was subsequently treated with the 5E-2N/SWNT composite healing-agent and thermally repaired for 15 minutes at 600 Celsius. The epoxy system restored its structure after 30 minutes, despite the fact that its mechanical healing effectiveness was not quantifiable. Similarly, 2.50% carbon nanotubes were combined with ethyl-phenylacetate and employed as therapeutic-agent in a capsule-based strategy [46]. When CNTs are combined with healing agents, they increase not only mechanical but electrical healing capacity. As a result, carbon nanotubes have been thoroughly investigated in extrinsic self-healing materials using a variety of techniques. CNTs, due to their tubular form, can be employed as a nano-reservoir for healing-agents, as a healing-agent, and in micro-capsule based healable polymer nanocomposite to restore mechanical qualities that have been deteriorated. Aside from this research, carbon nanotubes been employed to make carbon nanotubes-based polymer composites with intrinsic healing capabilities, which will be explored in the next section.

8.5.8 Intrinsic self-healing using CNTs composites made of polymers

Carbon nanotubes are mixed into polymers to create healable composites that may be employed in a variety of applications that need increased fracture resistance, material lifetime and flexibility. Conducive healable polymers for robotics, shear stiffening materials for body armour, etc., are just a few of the major applications. The next sections go through all of these various applications of carbon nanotube-based polymers with intrinsic-healing abilities.

8.5.9 Healable-conductive polymer-composites with multiple functions

SH multifunctional conductivity may be utilised in electronical appliances to restore circuit conductance, avoid damages and increase their lifespan. The electrical conductivity and mechanical characteristics of carbon nanotubes are quite high. By incorporating carbon nanotubes into an elastomer system, a conductive elastomer with autonomic healing capacity was created using a nanocomposite of poly-2-hydroxyethyl methacrylate and SWNTs coupled by interactions [47]. In the sample, both mechanical and electrical healing capacities were determined. A power supply and an LED bulb were linked to the nano-composite sample. Due to a lack of connection, the sample was promptly split into two halves and the LED was turned off. The sample pieces, however, mended in 5 minutes when maintained near to each other under ambient settings, and the LED bulb switched on again. For all of the different types of samples, the electrical healing efficiency was determined and this was approximately 95.00% with varying concentrations of SWNTs. As a result, SWCNTs were used to create electrically conductance and structurally stable elastomer that might be used to make sophisticated sensing devices. Another composite made of hyper-branched-poly-amido-amine (HPAMAM) polymers coated with carbon nanotube films and wrapped as a sushi shape showed SH properties, with the perfect electric-conductance recovery of about 20 minutes (Figure 8.2). The initial sample was linked with a circuit, the light was on; after the detached, the light was off; when it was recombined, the light was on again; and when the resistance was measured before and after disconnection, the light was off.

Figure 8.2 shows the SH conductive composite's sushi-like construction structure restored structural integrity and conductivity. Because the CNT

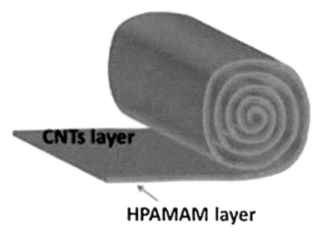

HPAMAM layer

Figure 8.2   Sushi-type structure shows the conductance HPAMAM/carbon nanotube composite disconnection's schematic structure. 2015 Copyright. The Royal Society of Chemistry has granted permission for this reprint.

layers are spirally coiled, the surfaces meet at the place of damage, restoring the conduction channels.

8.5.10 Self-healing polymer nanocomposites with shear-stiffening

The shear-thickening materials belong to the smart materials category. When the tension applied exceeds the critical shear rate, the viscosity of these materials has an unusual property: it dramatically rises. Due to their vital uses in military body armour, S-ST polymer composites have lately become a key study field. The most suited composites for high-performance body armour will be made by combining S-ST with other functions. MWCNTs are good nanofillers for reinforcing polymer materials because of their outstanding electrical conductivity, mechanical characteristics and low density. MWCNTs may be easily incorporated with the polymer matrix to generate multifunctional nanocomposites due to their superior dispersity than SWCNTs. Several studies have previously shown that adding MWCNTs into polymers improves the mechanical characteristics of the polymers. The MWNT-based S-ST polymer composite might be a promising body armour material with good protection and compression rate-dependent conductivity. Electrical self-healing has been shown in an MWNT/S-ST composite based on a polyboron dimethylsiloxane (PBDMS) derivative. The MWNT/S-ST-polymer composite was utilised to connect an LED light in a circuit on the other side. When the composite was divided into two parts, the LED shone brightly with a 9 V power supply before abruptly going out. When the fragmented pieces were brought together, the LED glowed brightly once again. This demonstrates the material's self-healing ability and the repaired specimen's little loss of conductivity. As a result, the MWCNT/S-ST polymer combination has a remarkable ability to self-heal at ambient temperature.

CNTs have an unusually high specific stiffness and strength, making them promising candidates for composite development. CNTs have strong electrical and thermal conductivities well as. CNTs have been widely investigated materials for the creation of diverse composite and smart materials due to these features, as well as their high aspect ratio, one-dimensional (1D) honeycomb lattice and low density [48]. The basic building component of graphite materials is graphene, which is a monolayer of carbon atoms firmly packed into a two-dimensional (2D) flat structure. The mechanical and electrical characteristics of open-ended CNTs are affected by the geometrical (2D) and electronic impacts of graphene on their field-emission properties, making them appropriate for the development of composite materials [49]. Furthermore, because of the overlap of 2pz orbitals of carbon atoms, graphene has the lowest energy and hence imparts anisotropic features to its composite materials. Owing to the variations in carbon atom bonding in-plane and out-of-plane, as well as their dimensional (3D) geometrical qualities graphene-based composites are possible filling agents because of their unique characteristics and cost-effectiveness.

Because of their low cost, low density, simple preparation pathways, diverse process capabilities and low thermal conductivity, carbon nanotube (CNT)/graphene-filled organic composites offer a lot of potential for creating cheaper thermoelectric materials for energy harvesting applications. In comparison to previously reported hybrid alloys, these features make them superior. Due to their distinctive structure and properties such as superconductivity, low weight, high stiffness and axial strength, CNT and graphene are now the most often utilised nanofillers.

8.5.11 Carbon nanotubes with customised shapes produce energy-collecting textile

There is a slew of studies looking into the cloth as a feasible method of energy harvesting. Fabric-based harvesting looks to offer a fruitful prospective region for such exploitation; thus, it makes sense. Carbon nanotubes (CNTs), which have previously exhibited several unique and valuable features in this field, are one substance employed to achieve this aim.

Rice University researchers worked with Tokyo Metropolitan University to develop a fibre-enhanced, flexible cotton fabric that uses carbon nanotubes as a thermoelectric (TE) energy source. It uses the well-known Seebeck phenomenon to convert heat into enough energy to light an LED. Using carbon nanotubes obviously has its drawbacks. Nonetheless, due to their one-dimensionality and unique features like flexibility and lightweight, they appear to be attractive possibilities. However, due to poor sample morphology and a lack of suitable Fermi energy tuning, sustaining the enormous power factor of individual carbon nanotubes in macroscopic assemblies has proven difficult [50–54].

8.6 Energy Collecting Modes

The different techniques of energy harvesting, as well as the several venues where such harvesting is required, include the following.

8.6.1 Energy harvesting for fossil fuel alternatives

With the Paris Agreement, a majority of world leaders have joined around a commitment to take a stronger stand against the threat of climate change. Each country is required to reduce emissions in order to keep global average temperatures from rising by more than $2°C$.

The time is ticking, and more new clean energy options must be created if this target is to be realised. The new economy examines some of the most intriguing sources of green energy currently being developed.

8.6.2 Elephant grass energy harvesting

Before coal, oil and gas were more easily available, biomass energy was the preferred fuel for millennia. CO_2 emissions are wreaking havoc on our environment today, and it is once again a significant contender in the global energy mix.

Biomass is any biological substance derived from plants or animals, however, it is most commonly in the form of wood. Next fuel, a clean-tech start-up based in Sweden, has developed a system that uses an alternative to wood pellets. Next Fuel offers a CO_2-negative alternative to fossil fuels that may be utilised directly in current energy infrastructure using elephant grass.

Elephant grass is a one-of-a-kind plant that can grow up to four metres in 100 days and yield many harvests each year. Following the collection of the grass, next fuel's technology takes relatively minimal energy to convert it into briquettes throughout the production process. The whole carbon cycle turns negative on an annual basis because less CO_2 is released into the environment when the fuel is burnt than was collected from the atmosphere a few months earlier while the grass was growing.

8.6.3 Energy harvesting hydrogen fuel cells

Hydrogen is one of the most plentiful elements on the planet, and while its usage in the power industry is not new, exciting new advancements have re-ignited interest in it. Hydrogen fuel cells can generate clean energy from a wide range of sources. They can be utilised in the transportation industry in the same manner as lithium-ion batteries can, but they do not need to be recharged.

1. Hydrogen is a renewable energy carrier that may be utilised to power cars or generate electricity. Only water and heat are produced when hydrogen is consumed.
2. To obtain pure hydrogen, we must first separate it from a combination.
3. Electrolysis is a technique for separating a water molecule into its oxygen and hydrogen atoms.
4. An electrical current is created when electrons are compelled to move across a circuit. After passing through the circuit, the electrons mix with hydrogen protons and oxygen molecules to form water (H_2O) and heat.
5. Because hydrogen fuel cells emit no carbon dioxide or other pollutants, they may be utilised to assist in the development of a zero-emission energy system.
6. Fuel cells may be used to power buildings, aeroplanes, cars and a variety of other electrical devices. Many fuel cells can be joined to make a fuel cell stack to supply the massive amounts of electricity required to power cars and other electronic devices.
7. Hydrogen enters the fuel cell through the anode electrode. Electrons (negative charge) and protons (positive charge) are separated from hydrogen atoms (positive charge).
8. The positively charged protons inside the fuel cell can travel through the membrane. Negatively charged electrons are unable to flow through the membrane and must instead go via a circuit.
9. Electrical current is created by electrons moving across a circuit.

Figure 8.3 Schematic diagram of the hydrogen fuel cell.

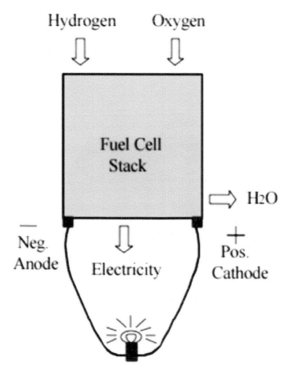

Figure 8.4 Fuel cell diagram.

Due to the high expense of hydrogen technology, Germany became the first country in the world to run passenger trains powered by hydrogen fuel cells. Reuters stated that European Union energy ministers decided to collaborate to boost hydrogen's prospects in the transportation and power sectors.

8.6.4 Solar paint as a source of energy

Solar panels are one of the most widely used renewable alternatives to fossil fuels, but what if you could capture the sun's energy without having to worry about the panels' environmental impact?

Researchers from Australia's Royal Melbourne Institute of Technology may have discovered an answer after producing a paint that can generate energy. By combining titanium oxide, which is found in many wall paints, with synthetic molybdenum-sulphide, the material may absorb solar energy as well as moisture from the surrounding air. A simple addition of the new compound to a brick wall might turn it into an energy and fuel source. The procedure is straightforward: The paint is constructed of titanium oxide and

Figure 8.5 Solar paint was applied on the wall of the house (courtesy: open source Wikipedia).

synthetic molybdenum-sulphide, a novel chemical. The qualities of the silica gel packet that arrive with new pairs of sneakers and other goods to wick away moisture are similar to those of that mouthful.

Solar energy is absorbed by synthetic molybdenum-sulphide before it is divided into hydrogen and oxygen. The hydrogen may then be collected and used to power a home, a vehicle, a truck, a boat, or an ATV.

8.6.5 Energy harvesting from waves

Using the energy provided by the ocean's waves appears to be a simple concept. Unfortunately, in practise, it is considerably more difficult. Researchers have been working on an ideal design for years.

In 2008 and 2009, a small-scale 'wave snake' project functioned off the coast of Portugal, but the intellectual property was handed to Wave Energy Scotland, a state organisation when the Scottish firm that developed the technology went bankrupt.

Wave energy research is continuously ongoing. Lockheed Martin, the world's biggest engineering firm, revealed plans. In order to encourage

Figure 8.6   Energy harvesting through wave energy (courtesy: open source Wikipedia).

private investment, the EU is cooperating with Wave Generating Scotland on a project to develop open-source software for wave and tidal energy systems.

Other ocean energy harvesting systems such as tidal power, ocean thermal energy conversion (OTEC) and saline processes pale in comparison to wave energy harvesters. Because waves can be found all throughout the ocean, they are more suitable for energy harvesting than tides.

Whisky is used to gather energy.

8.6.6 Energy harvesting whisky

For ages, Scotland has produced whiskey, and distilleries have developed methods to cope with the waste created as a by-product, frequently selling it to farmers as cow fodder. The industry is now seeking for a more creative approach to deal with the four million tonnes of rubbish it generates each year. The Green Alliance estimated in 2015 that the market for whiskey waste by-products may be worth £140 million ($184 million).

Some distilleries are now turning to anaerobic digester facilities to generate biogas, which is then converted into steam energy to power their operations. Glendullan distillery, owned by drinks giant Diageo, generated 6,000 MW hours of thermal energy in its first year of operation, lowering fossil fuel consumption by a quarter. By-products from whisky may also be used to generate heat and power automobiles.

8.6.7 Vehicle energy harvesting system (VEHS)

Continuous traffic and high-power usage are two apparently unconnected realities of city life. The vehicle energy harvesting system (VEHS), on the other hand, has created a novel connection that makes use of traffic movement to boost electricity access.

The VEHS harvests pressure from any existing or new road using a specifically engineered overlay layer, which is then used to power a traffic turbine and create electricity. Within six months, the system may be completed and commissioned, allowing for faster access to power. Aside from better electricity access, the VEHS's simple design will enable local assembly and a purposeful transition to local manufacturing. This, we hope, will equip local communities with much-needed skills and jobs, as well as help to revitalise their faltering economies.

8.6.8 Energy harvesting from sustainable power supply

Since its beginnings, the world of wireless devices has grown at an exponential rate. However, the range of services provided by these devices is frequently restricted by the battery life. As a result, a self-sustaining power source would propel us ahead to fully utilise the potential of such gadgets. Miniaturisation is driving the present technology revolution, and as gadgets get smaller, less energy is required onboard. This has prompted researchers to consider if batteries may be supplemented with technologies that continually collect otherwise squandered energy from the environment. Energy harvesting or scavenging is the practise of turning otherwise wasted ambient energy into a usable form. Energy harvesting would supplement a wireless system's long-term power source and might even eliminate the need to replace and maintain batteries in unfriendly or difficult-to-reach locations. To mention a few, safety monitoring devices, structure-embedded microsensors and medical implants are examples. Energy harvesting is also good for the environment since it provides a 'battery-free' solution by catching energy from natural sources including vibrations, heat, light and water and converting it to electricity.

8.6.9 Harvesting mechanical energy

Mechanical energy sources include vibrations and noise from industrial machinery and equipment, transportation, fluid flow such as air motions, biological locomotion such as walking and in-body motion such as chest and heart movement. Inertial and kinematic energy extraction from a mechanical source is the two basic methods [54–56]. Inertial energy is captured by a mass's resistance to acceleration. In these systems, a spring-mass-damper system is connected to the base at a single point. The bulk of the base vibrates due to its inertia, and these vibrations can be converted into electrical energy. Harvesters that exploit this idea of inertia include cantilever beams, pendulums and magnetoelastic oscillators. In kinematic energy harvesting, the energy-collecting transducer is directly coupled to various portions of the source. The transducer distorts as a result of the relative motion between these components, which is then converted into electrical energy. Two examples are harvesting energy by bending a tyre wall to check pressure or flexing and extending limbs to power a mobile phone. Mechanical energy is now converted to electricity via electrostatic, electromagnetic and piezoelectric transduction processes [57–60].

8.7 Advance Applications and Technologies of Energy Harvesting

Several energy-collecting devices are now in use, with several novel ones on the horizon. Light, heat, vibration and radio frequency are the most frequent energy sources (RF).

The development of ultralow-energy microcontroller units has produced a vast and developing power harvesting industry. The energy harvesting produced low-power wireless sensors, which now appear to be everywhere. The ripple effect, on the other hand, will expand throughout the consumer, commercial and scientific sectors, resulting in novel applications. Whether building small battery gadgets.

8.7.1 Mobile phone

By harvesting energy from the environment, you can attain the low power consumption required to run a cell phone. Mobile phones advanced technologies of wireless analogue telephones to portable computers, such as website surfing, movies, games and email facilities, while demanding strong battery duration.

For at least the last 10 years, the most important electrical design objective has always been low power. Wireless sensor networks with low power

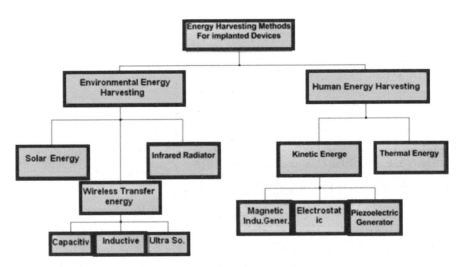

Figure 8.7 Applications of energy harvesting.

consumption are used in a variety of industrial, medical and commercial applications.

8.7.2 Solar power

Calculators, watches, toys, street light controllers, portable energy supplies and satellites are all examples of consumer and industrial applications for small solar cells. Because light sources are often sporadic, to offer a steady energy supply, solar power cells are employed on battery durabilities.

8.7.3 Thermoelectric

Thermoelectric harvesters rely on the Seebeck effect, which happens when two different metals have a temperature difference, resulting in a voltage. TEGs are made up of a series of thermocouples that are linked to a heat source such as a motor engine, a heater or even a solar panel.

8.7.4 Piezoelectric

When strained, piezoelectric transducers create electricity, making them ideal candidates for vibration sensors with energy-collecting modules that detect aircraft wing vibration and motor bearing noise. The cantilever creates an AC output voltage when it is moved by vibrations, which is rectified, controlled and as a battery base thin-film or supercapacitor.

8.8 Innovative Techniques and Technologies

In the coming years, some extremely exciting energy-harvesting labs might transform the future of the energy-harvesting sector.

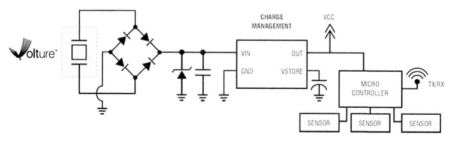

Figure 8.8   Midé Volture^{TM} piezoelectric energy harvester (courtesy: Midé).

8.8.1 Medical and fitness equipment

There are a few novel piezoelectric energy-collecting applications that are starting to emerge, which may then be used to operate a pacemaker or implanted, possibly removing the use for battery replacement. Other implanted devices are being studied to see whether they can be powered by body heat, movement or vibration.

- The patient sits on the chair with a low-frequency RF emitter, which the gadget receives, rectifies and stores.
- Gym-goers will be relieved to find that they can recoup part of the energy they waste there. Three British colleges have collaborated on the development of a piezo-electric power harvesting unit that connects to the knee and generates electricity when walking or running on a treadmill.

8.8.2 Antennas

On silicon and polyethylene substrates, NEC devices have been successfully prototyped but more funding and effort will be required to establish cost-effective mass manufacturing techniques. The researchers envision a system that works in tandem with traditional PV solar panels, harnessing hitherto untapped infrared energy.

8.9 Conclusion and Future Scope

Society is looking for materials that are crack- and damage-free in this era of polymer and necessary flexible electronics. In practise, industrial-scale production of self-healing materials will usher in a new era of technology, with the aerospace and military sectors in particular being dramatically transformed. This will extend the useful life of all man-made materials while simultaneously lowering maintenance costs and increasing flexibility. The addition of graphene and carbon nanotubes to polymer has resulted in the development of improved healing capacity. Graphene appears to be a potential SH material, but integrating it into polymer remains a difficult task. Graphene re-stacking is a key bottleneck in the manufacture of graphene-based polymer nanocomposites on a large scale. Similarly, carbon nanotubes have excellent conductance and healing characteristics, when coupled with polymers, it agglomerates. As a result, when employing various procedures to synthesise polymer nanocomposites, both graphene and carbon nanotubes confront significant limitations. A combination of CNT and graphene, on the other hand, may be able to address this. As a result, a polymer including a combination of CNT and graphene may have a greater healing efficiency and be healable by diverse stimuli such as IR, thermal, microwave, electrical and pH. The influence of CNTs and graphene in a polymer composite on intrinsic self-healing behaviour was investigated. Another extrinsic strategy is to examine the usage of CNT as a nanoreservoir in practise, as well as to undertake extensive self-healing studies. Although self-healing is still a pipe dream, we are not far from the day when man-made materials will be able to recover their structural integrity in the event of a breakdown.

Conductive polymer composites for energy harvesting conducting nanocomposite polymers are materials that undergo volume changes as a result of electrochemical the voltage loss caused by the electrolyte's resistance is called ohmic over potential, and the loss caused by the time it takes for ions to travel inside the electrolyte is called mass transport over potential. The resistance and hence the ohmic over potential would be reduced if the distance between the electrodes was reduced. By minimising the distance between electrodes, the time it takes for ions to go from one to the other is reduced, and the mass transport over potential is reduced. As a result, the electrode distance must be set to strike a balance between the two. The efficiency of conductive polymer composites in terms of mechanical-to-electrical energy conversion is quite poor. This is because the energy conversion from mechanical to electrical is mediated by a chemical reaction rather than being direct. As a result, they are ineffective as transducers in mechanical energy harvesters. Due to their cheap production cost, chemical and thermal stability and quick electron transfer kinetics, conductive polymers containing carbon nanotube fillers are attractive choices for constructing electrodes for thermogalvanic cells. To some extent, all processes involving energy conversion are inefficient. Motors, power transistors, automotive engines and light bulbs all grow heated and energy is squandered as heat in each instance. Radio stations produce megawatts of RF, although their signals are just microwatts when they reach antennas. Energy harvesting systems catch a portion of this squandered energy, convert it to power and put it to use.

Solar power panels and air turbines have become key energy sources. Large solar power panels and air turbines are the most well-known energy harvesting collectors. Energy harvesting systems, with careful design, can potentially replace batteries in some applications.

A system containing a controller for a chargeable Li-ion battery or thin film, a valve for the MCU or sensors and a wireless networking module must be carefully designed because the output of the energy-harvesting unit is typically intermittent. The closer an energy harvesting device can meet an embedded system's total demands, the closer that system may become battery-free.

References

[1] Agency Energy and environment (EE) report, 2008

[2] Commission IE Efficient electrical energy transmission and distribution, 2007

[3] T. M. Tritt, H. Böttner, L. Chen, Thermoelectrics: direct solar thermal energy conversion, MRS Bull. 33(4), 366–368, 2008.

[4] A. Abhat, Low temperature latent heat thermal energy storage: heat storage materials, Solar Energy, 30(4) 313–32, 1981.

[5] D. Buddhi, R L. Sawhney, In: Proceedings on thermal energy storage and energy conversion, 1994.

[6] M. Telkes, Thermal storage for solar heating and cooling. In: Proceedings of the workshop on solar energy storage sub-systems for heating and cooling of buildings, University of Virginia, Charlottesville, 1975.

[7] A. George, Hand book of thermal design, In: C. Guyer, editor. Phase change thermal storage materials. McGraw Hill Book Co., 1989.

[8] J. P. Heremans, C. M. Thrush, D. T. Morelli, M C. Wu, Thermoelectric power of bismuth nanocomposites. Phys. Rev. Lett. 88, 4361–5, 2002.

[9] W. He, G. Zhang, X. Zhang, J. Ji, G. Li, X. Zhao, Recent development and application of thermoelectric generator and cooler, Appl. Energy, 143, 1–25, 2015.

[10] J. Yang, T. Caillat, Thermoelectric materials for space and automotive power generation, MRS Bull. 31(3) 224–229, 2006.

[11] J. Changyoon, J. Chanwoo, L. Seonghwan, Q. F. Maria, B. P. Young, Carbon Nanocomposite Based Mechanical Sensing and Energy Harvesting, International Journal of Precision Engineering and Manufacturing-Green Technology, 1, 85, 2020

[12] A. Dey, O. P. Bajpai, A. K. Sikder, S. Chattopadhyay, M. A. S. Khan, Recent advances in CNT/graphene based thermoelectric polymer nanocomposite: A proficient move towards waste energy harvesting, Renewable and Sustainable Energy Reviews 53 653–671, 2016

[13] E. T. Thostenson, Ren Z, Chou T W, Advances in the science and technology of carbon nanotubes and their composites: a review. Compos. Sci. Technol. 61(13) 1899–1912, 2001.

[14] R. H. Baughman, A. A. Zakhidov, de Heer W A, Carbon nanotubes - the route toward applications. Science 297(5582) 787–792, 2002.

[15] J. Prasek, J. Drbohlavova, J. Chomoucka, J. Hubalek, O. Jasek, V. Adam, R. Kizek, Methods for carbon nanotubes synthesis – review, J Mater Chem 21(40) 15872–15884, 2011.

[16] S. Iijima, Helical microtubules of graphitic carbon. Nature 354(6348) 56–58, 1991.

[17] M. Farukh, R. Dhawan, B. P. Singh, S. Dhawan, Sandwich composites of polyurethane reinforced with poly (3,4-ethylene di oxythiophene)-coated multiwalled carbon nanotubes with exceptional electromagnetic interference shielding properties, RSC Adv 5(92) 75229–75238, 2015.

[18] R. Mathur, S. Pande, B Singh, T. Dhami, Electrical and mechanical properties of multiwalled carbon nanotubes reinforced PMMA and PS composites. Polym Compos 29(7) 717–727, 2008.

[19] A. P. Graham, G. S. Duesberg, W. Hoenlein, F. Kreupl, M. Liebau, R. Martin, B. Rajasekharan, W. Pamler, R. Seidel, W. Steinhoegl, E. Unger, How do carbon nanotubes fit into the semiconductor roadmap? Appl Phys A 80(6), 1141–1151, 2005.

[20] P. Jindal, S. Pande, P. Sharma, V. Mangla, A. Chaudhury, D. Patel, B. P. Singh, R. B Mathur, M. Goyal, High strain rate behavior of multi-walled carbon nanotubes-polycarbonate composites. Compos B Eng 45(1) 417–422, 2013.

[21] S. Pande, B. P Singh, R. B. Mathur, Processing and properties of carbon nanotube/polycarbonate composites, polymer nanotube nanocomposites: synthesis, properties, and applications, 2nd ed. Wiley, New Jersey, 333–364, 2014

[22] B. P. Singh, P. Saini, T. K. Gupta, P. Garg, G. Kumar, I. Pande, S. Pande, R. K Seth, S. K. Dhawan, R. B Mathur, Designing of multiwalled carbon nanotubes reinforced low density polyethylene nanocomposites for suppression of electromagnetic radiation, J Nanopart Res 13(12) 7065–7074, 2011

[23] F. C. Fim, N. R Basso, A. P. Graebin, D. S. Azambuja, G. B. Galland, Thermal, electrical, and mechanical properties of polyethylene–graphene nanocomposites obtained by in situpolymerization, J ApplPolym Sci 128(5) 2630–2637, 2013.

[24] B. Li, J. Zhang, Polysiloxane/multiwalled carbon nanotubes nanocomposites and their applications as ultrastable, healable and super hydrophobic coatings, Carbon 93, 648–658, 2015.

[25] Z. Wang, Y. Yang, R. Burtovyy, I. Luzinov, M. W. Urban, UV-induced self-repairing polydimethylsiloxane–polyurethane (PDMS–PUR) and polyethylene glycol–polyurethane (PEG–PUR) Cu-catalyzed networks, J Mater Chem A 2(37) 15527–15534, 2014.

[26] J. Ling, M. Z. Rong, M. Q. Zhang, Photo-stimulated self-healing polyurethane containing dihydroxyl coumarin derivatives, Polymer 53(13) 2691–2698, 2012.

[27] M. Verma, P. Verma, S. Dhawan, V. Choudhary, Tailored graphene based polyurethane composites for efficient electrostatic dissipation and electromagnetic interference shielding applications, RSC Adv 5(118) 97349–97358, 2015.

[28] P. Saini, V. Choudhary, B. Singh, R. Mathur, S. Dhawan, Enhanced microwave absorption behavior of polyaniline-CNT/polystyrene blend in 12.4–18.0 GHz range, Synth Met 161(15) 1522–1526, 2011.

[29] F. Shahzad, S. Yu, P. Kumar, J-W. Lee, Y-H. Kim, S. M. Hong, C. M. Koo, Sulfur doped graphene/polystyrene nanocomposites for electromagnetic interference shielding, Compos Struct 133, 1267–1275, 2015.

[30] A. Babal, R. Gupta, B. Singh, V. Singh, R. Mathur, S. Dhakate, Mechanical and electrical properties of high performance MWCNT/polycarbonate composites prepared by industrial viable twin screw extruder with back flow channel, RSC Adv 4, 64649–64658, 2014.

[31] G. Gedler, M. Antunes, J. Velasco, R. Ozisik, Enhanced electromagnetic interference shielding effectiveness of polycarbonate/graphene nanocomposites foamed via 1-step supercritical carbon dioxide process, Mater Des 90, 906–914, 2016.

[32] B. Shen, W. Zhai, M. Tao, D. Lu, W. Zheng, Enhanced interfacial interaction between polycarbonate and thermally reduced graphene induced by melt blending, Compos Sci Technol 86, 109–116, 2013.

[33] X. Zeng, J. Yang, W. Yuan, Preparation of a poly (methyl methacrylate)-reduced graphene oxide composite with enhanced properties by a solution blending method, Eur Polym J 48(10) 1674–1682, 2012.

[34] Y. M. Malinskii, V. Prokopenko, N. Ivanova, V. Kargin, Investigation of self-healing of cracks in polymers, Polym Mech 6(2), 240–244, 1970.

[35] R. P. Wool, Crack healing in semicrystalline polymers, block copolymers and filled elastomers. In: Adhesion and adsorption of polymers, Springer, Berlin, 341–362, 1980.

[36] R. Wool, K. O'connor, A theory crack healing in polymers, J Appl Phys 52(10), 5953–5963, 1981.

[37] S. R. White, N. Sottos, P. Geubelle, J. Moore, M. R. Kessler, S. Sriram, E. Brown, S. Viswanathan, Autonomic healing of polymer composites, Nature 409(6822) 794–797, 2001.

[38] M. Kessler, Self-healing: a new paradigm in materials design, proceedings of the institution of mechanical engineers, Part G. J AerospEng 221(4) 479–495, 2007

[39] G. Lanzara, Y. Yoon, H. Liu, S. Peng, W. Lee, Carbon nanotube reservoirs for self-healing materials, Nanotechnology 20(33), 335704, 2009.

[40] R. S. Sinha, D. Pelot, Z. Zhou, A. Rahman, X-F. Wu, A. L. Yarin, Encapsulation of self-healing materials by coelectrospinning, emulsion electrospinning, solution blowing and intercalation, J Mater Chem 22(18) 9138–9146, 2012.

[41] L. Mario, N. Ahmed, Organic thermoelectrics: Green energy from a blue polymer, Nat. Mater. 10(6) 409–410, 2011.

[42] B. Blaiszik, S. Kramer, S. Olugebefola, J. S. Moore, N. R. Sottos, White SR, Self-healing polymers and composites, Annu Rev Mater Res 40, 179–211, 2010

[43] Y. Yuan, T. Yin, M. Rong, M. Zhang, Self-healing in polymers and polymer composites. Concepts, realization and outlook: a review, Polym Lett 2(4) 238–250, 2008.

[44] F. Herbst, D. Döhler, P. Michael, W. H. Binder, Self-healing polymers via supramolecular forces, Macromol Rapid Commun 34(3), 203–220, 2013.

[45] B. Aissa, E. Haddad, W. Jamroz, S. Hassani, R. Farahani, P. Merle, Therriault D, Micromechanical characterization of single-walled carbon nanotube reinforced ethylidene norbornene nanocomposites for self-healing applications, Smart Mater Struct 21(10), 105028, 2012.

[46] B. M. Bailey, Y. Leterrier, S. Garcia, S. Van Der Zwaag, V. Michaud, Electrically conductive self-healing polymer composite coatings, Prog Org Coat 85, 189–198, 2015.

[47] W. Cui, J. Ji, Y-F. Cai, H. Li, R. Ran, Robust, Anti-fatigue, and self-healing graphene oxide/ hydrophobically associated composite hydrogels and their use as recyclable adsorbents for dye wastewater treatment, J Mater Chem A 3(33) 17445–17458, 2015.

[48] J. Jyoti, S. Basu, B. Singh, S. Dhakate, Superior mechanical and electrical properties of multiwall carbon nanotube reinforced acrylonitrile butadiene styrene high performance composites, Compos B Eng 83, 58–65, 2015

[49] S. Sharma, V. Gupta, R. Tandon, V. Sachdev, Synergic effect of graphene and MWCNT fillers on electromagnetic shielding properties of graphene–MWCNT/ABS nanocomposites, RSC Adv 6(22) 18257–18265, 2016.

[50] a) D. M. Rowe, in Handbook of electrics, CRC Press, Boca Raton, FL, 1995; b) D. Zhao, G. Tan, A review of thermoelectric cooling: materials, modeling and applications, Appl. Therm. Eng. 66(1–2) 15–24, 2014.

[51] a) S. B. Riffat, X. Ma, Thermoelectrics: a review of present and potential applications, Appl. Therm. Eng. 23(8) 913–935, 2003.

[52] F. J. Di Salvo, Thermoelectric cooling and power generation, Science 285(5428) 703–706, 1999.

[53] G. Chen, M. S. Dresselhaus, G. Dresselhaus, J.-P. Fleurial, T. Caillat, Recent developments in thermoelectric materials, Int. Mater. Rev. 48 (1) 45–66, 2003.

[54] G. J. Snyder, E. S. Toberer, Complex thermoelectric materials, Nat. Mater. 7(2) 105–114, 2008.

[55] B. C. Tee, C. Wang, R. Allen, Z. Bao, An electrically and mechanically self-healing composite with pressure-and flexion-sensitive properties for electronic skin applications, Nat Nanotechnol 7(12) 825–832, 2012.

[56] Y. Han, T. Wang, X. Gao, T. Li, Q. Zhang, Preparation of thermally reduced graphene oxide and the influence of its reduction temperature on the thermal, mechanical, flame retardant performances of PS nanocomposites, Compos A Appl Sci Manuf 84, 336–343, 2016.

[57] P. Garg, B. P. Singh, G. Kumar, T. Gupta, I. Pandey, R. Seth, R. Tandon, R. B. Mathur, Effect of dispersion conditions on the mechanical properties of multi-walled carbon nanotubes-based epoxy resin composites. J Polym Res 18(6) 1397–1407, 2010.

[58] B. P. Singh, K. Saini, V. Choudhary, S. Teotia, S. Pande, P. Saini, R. B. Mathur, Effect of length of carbon nanotubes on electromagnetic interference shielding and mechanical properties of their reinforced epoxy composites, J Nanopart Res 16(1) 1–11, 2014.

[59] L-C. Tang, Y-J. Wan, D. Yan, Y-B Pei, L. Zhao, Y-B. Li, L-B Wu, J-X. Jiang, G-Q. Lai, The effect of graphene dispersion on the mechanical properties of graphene/epoxy composites, Carbon 60, 16–27, 2013.

[60] R. B. Mathur, B. P. Singh, T. Dhami, Y. Kalra, N. Lal, R. Rao, A. M. Rao, Influence of carbon nanotube dispersion on the mechanical properties of phenolic resin composites,Polym Compos 31(2) 321–327, 2010

[61] Energy Information Administration, International Energy Outlook 2010. US: Technical Report, Department of Energy; 2010.

9

A PV Energy Harvesting System using MPPT Algorithms

Brian Azzopardi[1], Vibhu Jately[2], and Jyoti Joshi[3]

[1]MCAST Energy Research Group (MCAST Energy), Institute of
Engineering and Transport, Malta College of Arts,
Science and Technology (MCAST), Malta
[2]Department of Electrical and Electronics Engineering,
University of Petroleum and Energy Studies, India
[3]Department of Computer Science and Engineering,
Graphic Era Hill University, India
E-mail: Brian.Azzopardi@mcast.edu.mt; vibhujately@gmail.com;
jyotijoshi@gehu.ac.in

Abstract

Photovoltaic (PV) technology has seen tremendous growth within the last
few decades. For increasing the efficacy of solar PV harvesting system
maximum power point tracking (MPPT) control is invariably used in most
PV applications. The aim of MPPT algorithms is to assure that peak power
is extracted from the PV and supplied to the output side irrespective of
the changes in the atmospheric and load conditions. Among several MPPT
methods, hill climbing (HC) based MPPT algorithms have seen significant
attention because of their good tracking capability. This chapter compares
the behaviour of conventional and adaptive HC MPPT techniques in MAT-
LAB/Simulink environment. The algorithms are compared on the basis of
power oscillations during steady-state, tracking speed and deviation from
the MPP tracking path. The results indicate that there exists a trade-off
between conventional and adaptive MPPT techniques between the speed of
convergence and deviation from the tracking route. Although the adaptive
hill climbing algorithm has a higher tracking speed, it suffers from a large

deviation from the peak power tracking path when compared to conventional methods during the sudden dynamic variation in irradiance.

Keywords: PV Energy Harvesting; MPPT; Hill Climbing; Perturb and Observe; Incremental Conductance.

9.1 Introduction

The energy crisis and climate change act have forced mankind to look for new cleaner ways of energy production [1]. Among them, solar photovoltaic has seen a considerable amount of attention due to its wide range of benefits. Solar PV is one of the cleanest energy harvesting technologies and has a very low running cost as it converts solar energy into electricity using the photoelectric effect. Moreover, due to the mass production of PV modules, the cost of PV systems has significantly reduced in recent years. Evidently, the output from the PV depends on the point of operation on the non-linear I–V curve of the PV cell as shown in Figure 9.1. The maximum power transfer theorem states that the peak power is transferred from the input side to the output side when the input resistance is equal to the load resistance of the PV system. Hence, peak power is extracted from PV when the load line crosses the I–V characteristics at the knee point which is referred to as the MPP [2].

The peak power tracking method helps in constantly tracking the maximum power point of the I–V curve. However, the task of the MPPT algorithm becomes more tedious when the I–V characteristic changes under variations in environmental conditions as shown in Figure 9.2.

Various MPPT methods are present in the literature to efficiently locate the MPP of the PV module. In [3], the authors proposed a dual-tracking technique to increase the speed of convergence of the MPP tracking process. The technique increases the tracking speed by using two perturbation variables, voltage and current. In [4], the authors proposed a tunicate swarm algorithm-based particle swarm optimisation MPPT method to track the global maximum power point under partial shading conditions. The hybrid algorithm has a good exploration and exploitation capability which reduces the tracking time and avoids being trapped in the local maxima. In [5], the authors proposed an improved drift avoidance technique under simultaneous sudden changes in irradiance and load resistance. The method shows improved tracking capability, but its performance remains in doubt under small changes in load resistance.

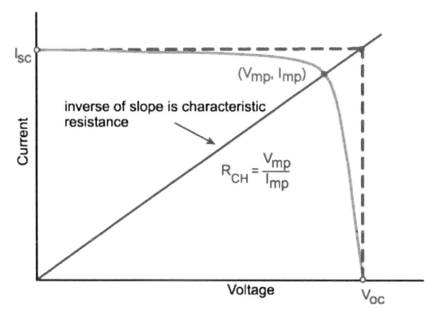

Figure 9.1 I–V curve of photovoltaic cell [3].

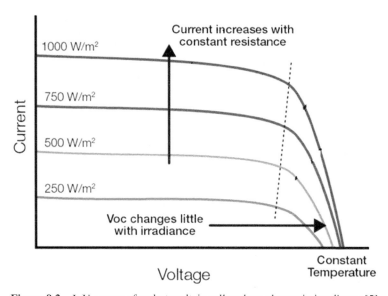

Figure 9.2 I–V curves of a photovoltaic cell under a change in irradiance [5].

In [6], the authors proposed a PI-based P&O technique to overcome the sluggish behaviour of the P&O algorithm. However, it is challenging to determine the gains of the controller for a wide range of irradiance. The authors in [7, 8] compared the behaviour of HC methods during sudden variations in irradiance levels. The investigation indicates that the resolution of ADC is a key attribute for the practical realisation of peak power point tracking methods. In [9], the authors proposed an adaptive MPPT technique. The speed of tracking is highly dependent on the value of the slope of the P–V curve which determines the steady-state boundary. Several authors have compared the behaviour of classical and modified HC methods [10–[12]]. However, there exists a research gap to evaluate the performance comparison between conventional and adaptive hill climbing algorithms based on the drift presence under a sudden change in irradiance.

9.2 Challenges in Hill Climbing MPPT Algorithms

The task of the peak power point algorithm is to quickly converge towards the peak power point with reduced steady-state oscillations. Among hill climbing techniques, P&O and INC are the two most used MPPT algorithms. As the name suggests, the P&O method provides a perturbation and observes the response of the system on the application of a perturbation. If the outcome is in favour, the next run goes without any change in the direction of the perturbation, otherwise, it goes with a perturbation in the reverse direction. The P&O algorithm observes the difference in power and voltage to evaluate whether to increase/decrease the duty cycle of the converter. On the other hand, the INC algorithm observes the slope of the I–V characteristic to deduce the sign of the subsequent perturbation. The conventional hill climbing algorithms operate with a fixed perturbation step size. The small step size reduces the power ripples but also reduces the speed of convergence, whereas using a large step size accelerates the speed of convergence with large power ripples as shown in Figure 9.3. On the other hand, the adaptive hill climbing algorithms adaptively change the step size depending upon whether the point of operation is near or away from the peak power point. Another important drawback in classical and adaptive HC methods is that they tend to drift away from the peak power point tracking path under varying irradiance. This is due to the lack of information present within the algorithm to evaluate the logic for the change in power, i.e., the change is because of intentional perturbation or variation in irradiance. This leads to the point of operation to deviate from the correct tracking path. The deviation is severe if the step size chosen is large. The next section discusses the commonly used conventional and adaptive HC techniques.

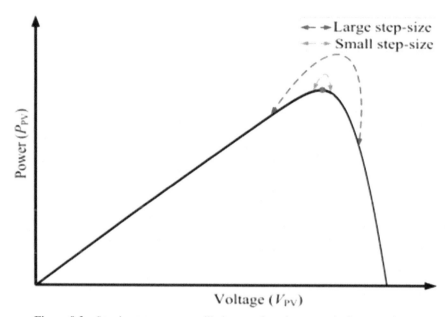

Figure 9.3 Steady-state power oscillations are based on perturbation step size [7].

9.3 HC MPPT Techniques

Some commonly used HC peak power point tracking techniques along with their adaptive versions are given below:

9.3.1 Conventional perturb and observe (P&O) method

The P&O technique is the most popular method for peak power point tracking in the PV industry. The algorithm operates by measuring the voltage and current of the PV array. The voltages between two-time stamps are measured to determine whether the converter's duty cycle needs a change. Similarly, the powers between two-time stamps are measured to determine the change in power. The algorithm measures the change in power to determine whether the power has increased or decreased. After that, the algorithm measures the change in voltage to determine whether the voltage has increased or decreased. If the power has increased due to the increment in the voltage, the voltage is increased by changing the duty to further raise the power. On the other hand, if the power has decreased due to the increment in the voltage, the voltage is reduced by changing the duty to increase the power. The value of increment or decrement in the duty determines the tracking speed and power ripples around MPP, as previously discussed.

9.3.2 Adaptive P&O method

To overcome the dilemma of speedy convergence vis-à-vis large power oscillations at steady-state, adaptive perturb and observe algorithm adaptively changes the tracking step size. If the operating point is in the transient state, that is d*P*/d*V* is large, a large perturbation steps-size is used to track the peak power point. As the operating point reaches steady-state the value of d*P*/d*V* decreases, requiring in turn a reduction in the step size to ensure low power oscillations around MPP. The flowchart of the conventional and adaptive perturb and observe algorithms is shown in Figure 9.4. Another important variable 'M' helps in good tracking speed and low steady-state oscillations should be carefully selected. The value of M chosen for the study is 0.1. This is calculated by determining the maximum change in voltage, the maximum difference in duty cycle and the maximum difference in power between two perturbations as given in eqn. (9.1):

$$M = \frac{|dV_{max}| \times \Delta D_{max}}{|dP_{max}|}. \qquad (9.1)$$

In Figure 9.4, the absolute value of the gradient of P–V characteristic is used for determining the step size. This is done to accurately determine the sign of the step size regardless of the operating point being on the LHS or RHS of the power curve. It is interesting to note that conventional and adaptive P&O algorithms cannot determine variations in irradiance and drift away from the peak power tracking path.

9.3.3 Conventional incremental conductance (INC) method

Incremental conductance is another famous MPPT method belonging to the class of hill climbing techniques. The INC algorithm, as discussed already, uses the information of the I–V curve and measures the difference in current and voltage between two successive time stamps. As its name indicates, the INC algorithm compares the present conductance with the incremental conductance to determine whether the operating point is on the LHS or

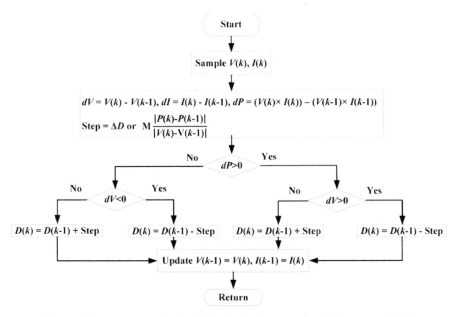

Figure 9.4 Flowchart of classical and adaptive perturb and observe method.

RHS of the I–V characteristic. If the algorithm finds that the operating point lies on the LHS of the curve the PV voltage is incremented to reach the peak power point. On the contrary, the PV voltage is decremented if the operating point lies on the RHS of the I–V curve. The reference value of voltage is incremented or decremented using a fixed perturbation step. Since the perturbation variable is voltage a meticulously designed PI controller is employed to arrive at the appropriate duty cycle to reach the target voltage. For stable operation, it has to be ensured that the perturbation is applied after the voltage, current and power values have reached a steady state.

9.3.4 Adaptive INC method

The adaptive INC algorithm quickly adapts the perturbation step to speed up convergence and reduce steady-state power oscillations. Using an approach, similar to the adaptive P&O method, the step here is adjusted in accordance with the absolute value of the gradient of the P–V curve, that is dP/dV. The adaptive INC technique incorporates a scaling factor 'N' which is dependent on the maximum change in voltage step size, maximum deviation in voltage and power between two-time stamps as in eqn. (9.2). The flowchart of the conventional and adaptive INC algorithm is shown in Figure 9.5. Taking the

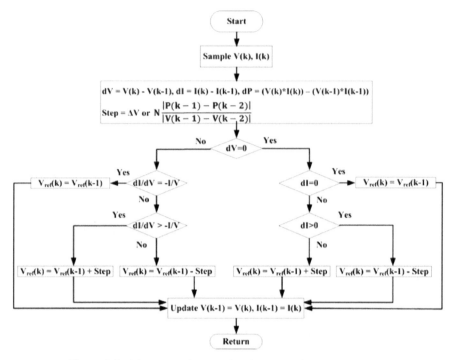

Figure 9.5   Flowchart of conventional and adaptive INC algorithm.

absolute value of the gradient of the power curve ensures that only the dP/dV is considered without the sign to calculate the perturbation step size. After each iteration, the present time stamp values are updated and stored as the previous iteration value:

$$N = \frac{|dV_{max}|\Delta V_{max}}{|dP_{max}|}. \tag{9.2}$$

9.4 Design of Boost Converter for Impedance Matching

The above-discussed MPPT algorithms are implemented using a step-up converter, illustrated in Figure 9.6. There are two operating modes for the step-up converter. In mode 1, the switch is turned on at $t = 0$ and switched off at $t = t_{on}$. As soon as the switch is turned 'ON' the inductor current which rises linearly flows through the inductor L and the switch SW. In mode 2, the switch is turned off at $t = t_{on}$ and then again turned on at $t = T_S$. In this

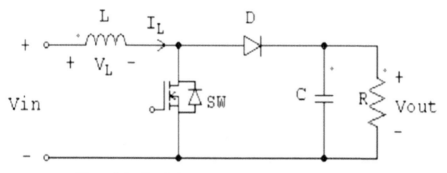

Figure 9.6 Circuit diagram of the DC–DC boost converter.

mode, by flipping the switch off, a linearly decreasing current flows through L, diode D, capacitor and thereby energy stored is dumped to load R. Hence, the energy stored in L is shifted to the load side. This process boosts the input-side voltage to an increased value and can be calculated using eqn. (9.3):

$$V_{out} = \frac{1}{(1-d)}V_{in},\tag{9.3}$$

where V_{out} is the output voltage, d is the duty and V_{in} is the input-side voltage. From eqn. (9.3), we can observe that V_{out} is inversely proportional to $(1-d)$. Hence, if d is equal to 1, there will be no transfer of energy. In continuous conduction mode, I_L never becomes zero for a minimum value of inductance which can be calculated using eqn. (9.4):

$$L = \frac{V_{in} \times (V_{out} - V_{in})}{\Delta I_L \times F_s \times V_{out}},\tag{9.4}$$

where F_S is the switching frequency of converter and ΔI_L is inductor ripple current expressed in eqn. (9.5):

$$\Delta I_L = 0.2 \times I_{out(max)} \times \frac{V_{out}}{V_{in}},\tag{9.5}$$

where *I*out(max) is the maximum output-side current. The capacitance for desired output-side voltage ripple is calculated using eqn. (9.6):

$$C = \frac{I_{out(max)} \times d}{F_s \times \Delta V_{out}},\tag{9.6}$$

where ΔV_{out} is the desired output voltage ripple as in eqn. (9.7).

$$\Delta V_{out} = ESR \times (\frac{I_{out(max)}}{1-d} + \frac{\Delta I_L}{2}),\tag{9.7}$$

where ESR is equivalent series resistance of output capacitor.

9.5 Results and Discussion

The behaviour of MPPT algorithms is tested on an 80 W PV module in MATLAB/Simulink environment as in Figure 9.7. The rating of the 80-W PV module is given in Table 9.1. The response of the MPPT algorithms has been carried out to effectively assess their working under dynamic change in irradiance.

To demonstrate the behaviour of the MPPT algorithms under dynamic change in irradiance, the peak power point tracking methods are tested with an initial irradiance of $G = 1000$ W/m^2 and temperature at $T = 25°C$. At $t = 0.2$ s, G is gradually decreased to $G = 500$ W/m^2 till $t = 0.3$ s. The irradiance is then linearly increased back to $G = 1000$ W/m^2 up to $t = 0.4$ s as shown in Figure 9.8 (a). The power response curves of classical and adaptive P&O techniques are shown in Figure 9.8 (b) and (c), respectively. The response curve indicates that the conventional P&O has a small deviation with respect to adaptive P&O algorithm. This is due to large change in step proportionate to large variation in irradiance. Similarly, the response curves of conventional and adaptive INC algorithm indicate the same behaviour as in Figure 9.8 (d) and (e), respectively. Hence, overall efficiency of adaptive P&O and INC method gets reduced due to large deviation from the MPP tracking path.

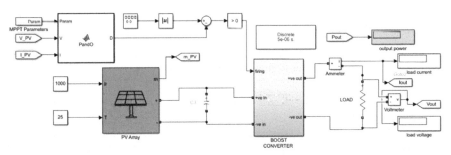

Figure 9.7   MATLAB/Simulink model of solar PV energy harvesting system.

Table 9.1   Specifications of the 80 W PV module.

N_c	V_{OC}	I_{SC}	V_{MPP}	I_{MPP}	β_{OC}	α_{SC}
72	21.95 V	4.88 A	17.84 V	4.5 A	−0.31 V/°C	0.058 A/°C

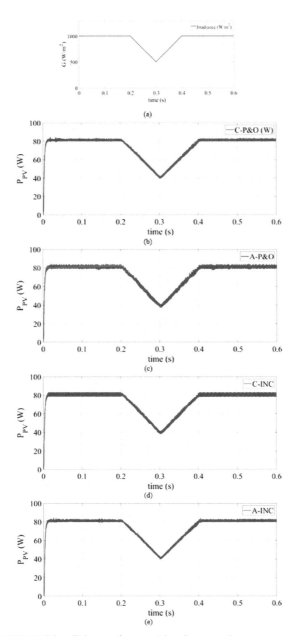

Figure 9.8 MATLAB/Simulink waveforms (a) irradiance and power curves of (b) classical P&O, (c) adaptive P&O, (d) classical INC and (e) adaptive INC MPPT algorithm.

9.6 Conclusion

This chapter has discussed the behaviour of solar PV energy harvesting system during dynamic change in irradiance. Although, hill climbing algorithms have a good MPP tracking capability, the adaptive type P&O and INC MPPT techniques suffer from large deviations from peak power point tracking path during dynamic varying irradiance. Hence, the modifications in the hill climbing algorithms are required to accurately detect the variation in irradiance. The scaling factor that adaptively changes the step size should be carefully evaluated for a broad scope in irradiance levels.

References

[1] S. K. Nag, T. K. Gangopadhyay and J. Paserba, "Solar Photovoltaics: A Brief History of Technologies [History]," in IEEE Power and Energy Magazine, vol. 20, no. 3, pp. 77-85, May-June 2022, doi: 10.1109/MPE.2022.3150814.

[2] N. Swaminathan, N. Lakshminarasamma and Y. Cao, "A Fixed Zone Perturb and Observe MPPT Technique for a Standalone Distributed PV System," in IEEE Journal of Emerging and Selected Topics in Power Electronics, vol. 10, no. 1, pp. 361-374, Feb. 2022, doi: 10.1109/JESTPE.2021.3065916.

[3] V. Jately and S. Arora, "Development of a dual-tracking technique for extracting maximum power from PV systems under rapidly changing environmental conditions," Energy (Oxf.), vol. 133, pp. 557–571, 2017.

[4] A. Sharma, A. Sharma, V. Jately, M. Averbukh, S. Rajput, and B. Azzopardi, "A Novel TSA-PSO Based Hybrid Algorithm for GMPP Tracking under Partial Shading Conditions," Energies, vol. 15, no. 9, p. 3164, Apr. 2022, doi: 10.3390/en15093164.

[5] V. Jately, S. Bhattacharya, B. Azzopardi, A. Montgareuil, J. Joshi and S. Arora, "Voltage and Current Reference Based MPPT Under Rapidly Changing Irradiance and Load Resistance," in IEEE Transactions on Energy Conversion, vol. 36, no. 3, pp. 2297-2309, Sept. 2021, doi: 10.1109/TEC.2021.3058454.

[6] A. K. Abdelsalam, A. M. Massoud, S. Ahmed and P. N. Enjeti, "High-Performance Adaptive Perturb and Observe MPPT Technique for Photovoltaic-Based Microgrids," in IEEE Transactions on Power Electronics, vol. 26, no. 4, pp. 1010-1021, April 2011, doi: 10.1109/TPEL.2011.2106221.

[7] V. Jately, B. Azzopardi, J. Joshi, B. Venkateswaran V, A. Sharma, and S. Arora, "Experimental Analysis of hill-climbing MPPT algorithms under low irradiance levels," Renewable and Sustainable Energy Reviews, vol. 150, p. 111467, Oct. 2021, doi: 10.1016/j.rser.2021.111467.

[8] V. Jately and S. Arora, "Performance Investigation of Hill-Climbing MPPT Techniques for PV Systems Under Rapidly Changing Environment," Advances in Intelligent Systems and Computing, pp. 1145–1157, 2018, doi: 10.1007/978-981-10-5903-2_120.

[9] J. Ahmed and Z. Salam, "A Modified P&O Maximum Power Point Tracking Method With Reduced Steady-State Oscillation and Improved Tracking Efficiency," in IEEE Transactions on Sustainable Energy, vol. 7, no. 4, pp. 1506-1515, Oct. 2016, doi: 10.1109/TSTE.2016.2568043.

[10] S. B. Kjær, "Evaluation of the "Hill Climbing" and the "Incremental Conductance" Maximum Power Point Trackers for Photovoltaic Power Systems," in IEEE Transactions on Energy Conversion, vol. 27, no. 4, pp. 922-929, Dec. 2012, doi: 10.1109/TEC.2012.2218816.

[11] J. A. Carrasco, F. G. de Quirós, H. Alavés and M. Navalón, "An Analog Maximum Power Point Tracker With Pulsewidth Modulator Multiplication for a Solar Array Regulator," in IEEE Transactions on Power Electronics, vol. 34, no. 9, pp. 8808-8815, Sept. 2019, doi: 10.1109/TPEL.2018.2886887.

[12] V. Jately and S. Arora, "An efficient hill-climbing technique for peak power tracking of photovoltaic systems," 2016 IEEE 7th Power India International Conference (PIICON), 2016, pp. 1-5, doi: 10.1109/POW-ERI.2016.8077327.

Index

About the Editors

Shailendra Rajput is an Associate Professor at Xi'an International University, China. He was a postdoctoral fellow at Ariel University, Israel, Xi'an Jiaotong University, China and Indian Institute of Technology, Kanpur, India. Dr. Rajput is also affiliated with Ariel University, Israel as Research Fellow. Dr. Rajput received the B.Sc. and M.Sc. degrees from the Dr. Hari Singh Gour University, Sagar, India in 2006 and 2008, respectively. Dr. Rajput received the Ph.D. degree from the Birla Institute of Technology, Ranchi, India in 2014. His main research work is associated with Energy harvesting, Solar energy, Energy storage, Energy materials, Ferroelectricity, Piezoelectricity, and biomedical application of electromagnetic waves.

Abhishek Sharma received a bachelor's degree in electronics and communication engineering from ITM-Gwalior, India, in 2012, and a master's degree in robotics engineering from the University of Petroleum and Energy Studies (UPES), Dehradun, India, in 2014. He was a Senior Research Fellow in a DST-funded project under the Technology Systems Development Scheme and worked as an Assistant Professor with the Department of Electronics and Instrumentation, UPES. He also worked as a research fellow at Ariel University (Israel) and received the Emerging Scientist award in 2021. Currently he is working as a assistant professor in computer science department at Graphic Era University and as a adjunct lecturer in UCSI University, Malaysia. His research interests include machine learning, optimization theory, swarm intelligence, embedded system, control and robotics.

Vibhu Jately received his Ph.D. degree from G. B. Pant University, Pantnagar, India. Following that, he worked as an Assistant Professor under United Nations Development Program within the Department of Electrical and Computer Engineering at Wollo University, Ethiopia. After that he worked as a Post-Doctoral Research Fellow for two years at MCAST Energy Research Group, Malta where he was a Task Leader of European H2020 projects. Currently he is working as an Assistant Professor (Selection Grade) within the Department of Electrical & Electronics Engineering at the University of Petroleum & Energy Studies, Dehradun, India. He has over 8 years of teaching and research experience. His area of interest includes power electronics applications in renewable energy systems and has worked in formulating MPPT algorithms, control strategies in grid integration of PVs, microgrids and optimization algorithms in PV applications. He is an active researcher and has published several research articles in top-quality peer-reviewed journals and international conferences.

Mangey Ram received his PhD, major in Mathematics and minor in Computer Science from G. B. Pant University of Agriculture and Technology, Pantnagar in 2008. He is an editorial board member in many international journals. He has published 102 research publications in national and international journals of repute. His fields of research are Operations Research, Reliability Theory, Fuzzy Reliability and System Engineering. Currently, he is working as a Professor at Graphic Era University.